Particle and Nanoparticle Functionalization 2018

Topical at the 8th World Congress on Particle Technology

Orlando, Florida, USA
22 - 26 April 2018

ISBN: 978-1-5108-6977-6

Printed from e-media with permission by:

Curran Associates, Inc.
57 Morehouse Lane
Red Hook, NY 12571

Some format issues inherent in the e-media version may also appear in this print version.

Printed by Curran Associates, Inc. (2018)

For permission requests, please contact AIChE
at the address below.

AIChE
120 Wall Street, FL 23
New York, NY 10005-4020

Phone: (800) 242-4363
Fax: (203) 775-5177

www.aiche.org

Additional copies of this publication are available from:

Curran Associates, Inc.
57 Morehouse Lane
Red Hook, NY 12571 USA
Phone: 845-758-0400
Fax: 845-758-2633
Email: curran@proceedings.com
Web: www.proceedings.com

TABLE OF CONTENTS

Some abstracts in this book refer to images that were not made available by the publisher

511465 UV Light-Inhibited Release for Time Controlled Growth Factor Delivery from TiO_2 Nanoparticles in a Molecular Hydrogel (Invited)

Abstract:

Tissue engineering scaffold materials mimic the natural extracellular matrix (ECM) to promote tissue regeneration and degrade naturally as healthy tissue grows. An ideal growth factor delivery system for regenerative medicine should operate synergistically with these tissue engineering materials. Spatial control of growth factor delivery is achieved by loading them into an implanted or injected scaffold, which acts as a reservoir for diffusional delivery to the surrounding tissue. The goal is often to provide a consistent therapeutic concentration through sustained, ongoing delivery, although depending on the application, many different delivery profiles can be required, such as rapid burst or pulsatile release. Advancement of the field depends on building a library of control techniques so as to be able to engineer any possible combination of different release profiles of multiple drugs or growth factors into a single material.

We have previously used chemical cross-linking via SMCC to provide a permanent ongoing presentation of growth factors by immobilising the growth factors to electrospun nanofibres in tissue engineering scaffold materials[1-3]. We have also used this SMCC attachment to temporally control their non-permanent delivery, by attaching polysaccharide chains to the growth factor to delay their release from self-assembling peptide (SAP) hydrogel tissue engineering materials[4].

Here, we have used SMCC attachment method and nanoparticle delivery to explore two new mechanisms for temporal control of delivery: attachment to a carrier nanoparticle embedded within a SAP hydrogel, and stimuli-responsive control of delivery from these nanoparticles. We report on a system in which nanoparticle bound BDNF shows greater release from the SAP hydrogel before UV treatment, at which point the unbound growth factor adsorbs to the SAP nanofibres. We also use this relationship and the UV-sensitive nanoparticles to induce a consistent release profile, using the UV response to counter the natural non-zero order delivery profile.

REFERENCES:

1. Wang T-Y, Bruggeman KAF, Sheean RK, Turner BJ, Nisbet DR, Parish CL. Characterization of the stability and bio-functionality of the tethered proteins on bioengineered scaffolds: implications for stem cell biology and tissue repair. J Biol Chem. 2014;289:15044-15051.
2. Potas JR, Haque F, Maclean FL, Nisbet DR. Interleukin-10 conjugated electrospun polycaprolactone (PCL) nanofibre scaffolds for promoting alternatively activated (M2) macrophages around the peripheral nerve in vivo. J Immunol Methods. 2015;420:38-49.
3. Horne MK, Nisbet DR, Forsythe JS, Parish CL. Three-dimensional nanofibrous scaffolds incorporating immobilized BDNF promote proliferation and differentiation of cortical neural stem cells. Stem Cells Dev. 2010;19(6):843-852.
4. Bruggeman KF, Rodriguez AL, Parish CL, Williams RJ, Nisbet DR. Temporally controlled release of multiple growth factors from a self-assembling peptide hydrogel. Nanotechnology. 2016;27(38):385102.

508525 Investigating Functionalized Particle – Microbe Interactions on 'Touch' Surfaces

Abstract:

Abstract: Disinfection of surfaces play a major role in deterring contact based pathogen transfer, thereby reducing infection rates. Common disinfection approaches employ the use soap and water (highly effective but not strong enough) and antimicrobial actives which are often harsh to the substrate/surfaces, and ultimately lead to antimicrobial resistance development. The primary objective of this study is to investigate a 'non-kill' based alternate disinfection approach focused on enhanced removal of microbes from skin via particle - bacteria interaction and stripping. The hypothesis that functionalized particles capable of interacting strongly with bacteria can be used to strip bacteria and disinfect surfaces more efficiently, has been verified with a set of functionalized particles for removing *E. coli* from a polymeric substrate. Particle momentum, particle – bacteria interaction strength, and particle concentration determined the disinfection efficacy. It was determined that modified particles could reduce bacterial concentrations to below detection limits (>4 log reduction) without inducing bacterial kill.

509029 Green Synthesis of Fluorescent Nanomaterials for Optical Bioimaging and Beyond

Abstract:

Fluorescent nanoparticles have received intense scientific attention and offer promising applications in many fields from optoelectronic devices to cancer treatment. The development of robust mass-scale nanoparticle synthesis methods is one of key materials challenges in moving towards fluorescent nanoparticles for commercialization and advanced applications. In this talk, I will summarize our work on green and scalable synthesis of fluorescent nanoparticles for multifunctional applications, including biomass-derived carbon dots for sensing and bioimaging, subcritical water nanonization of organic compounds for diagnosis and therapy, preparation of lanthanide-doped inorganic nanoparticles by high gravity method for transparent, flexible and luminescent composite films. A brief overview of this exciting field, along with some challenges and opportunities, will also be presented.

515286 Superparamagnetic Nanoparticles for Triggered Drug Release from Alginate Hydrogels

Abstract:

While magnetic nanomaterials have already been used in clinics for contrast enhancement in magnetic resonance imaging (MRI) (Singh and Sahoo, 2014), there has been no clinical approval for a drug delivery system containing such nanoparticles, yet. Magnetic nanoparticles currently investigated for their possible application in biomedicine are predominantly different crystalline polymorphs of iron oxides. For small enough crystal sizes, iron oxides exhibit the so-called "superparamagnetic" behavior, a feature combining high magnetization with very low coercive forces. Such superparamagnetic nanoparticles show great potential in therapeutic applications due to their ability to transform the energy of an alternating magnetic field to thermal energy in what is often called magnetic fluid hyperthermia (Jordan *et al*, 1999). These particles dissipate thermal energy by magnetic relaxation through the Brownian and Neel mechanisms. Therefore, aqueous suspensions at relatively higher nanoparticle concentrations (g/L), the so-called "ferrofluids", also increase their temperature in the presence of an alternating magnetic field (Sotiriou *et al*, 2013).

In this work, a composite multi-scale structure consisting of the biopolymer alginate, functional nanoparticles and a model drug is fabricated and analyzed. We demonstrate the highly scalable and reproducible synthesis of uniformly SiO_2-coated superparamagnetic Fe_2O_3 nanoparticles (Teleki *et al*, 2009), and evaluate their suitability as stimuli-responsive nanofillers in a drug-loaded biopolymer alginate matrix. The superior colloidal stability of the SiO_2-coated Fe_2O_3 nanoparticles over their uncoated counterparts and their dispersibility in aqueous suspensions facilitates their incorporation in alginate hydrogel microbeads. We perform detailed physicochemical and magnetic characterization on the hybrid alginate hydrogel beads and evaluate their potential in magnetic fluid hyperthermia and enhanced biomolecule release in the presence of an external alternating magnetic fields. We examine the hyperthermia performance of such multiscale particle structures in the presence of alternating magnetic fields and compare the release of dextran (a model biomolecule) in the presence and absence of external stimuli. The enhanced triggered release of dextran in the presence of magnetic fields further highlights the potential of such superparamagnetic SiO_2-coated Fe_2O_3 nanoparticles as a functional transducer in such systems. The possibility to externally stimulate drug release will open up new possibilities in intelligent, on-demand drug administration (Teleki *et al*, 2016).

Jordan A., R. Scholz, P. Wust, H. Fahling, R. Felix, *J. Magn. Magn. Mater.*, **201**, 413-419 (1999).

Singh, A., S. K. Sahoo, *Drug Discov. Today*, **19**, 474-481 (2014).

Sotiriou G.A., M. A. Visbal-Onufrak, A. Teleki, E. J. Juan, A. M. Hirt, S. E. Pratsinis, C. Rinaldi, *Chem. Mater.*, **25**, 4603-4612 (2013).

Teleki A., M. Suter, P. R. Kidambi, O. Ergeneman, F. Krumeich, B. J. Nelson, S. E. Pratsinis, *Chem. Mater.*, **21**, 2094-2100 (2009).

Teleki A., F. L. Haufe, A. M. Hirt, S. E. Pratsinis, G. A. Sotiriou, *RSC Adv.* **6**, 21503-21510 (2016).

508350 The Silanol Content of Silica Nanoparticles (Invited)

Abstract:

The surface chemistry of synthetic amorphous silicas is important for their performance (*e.g.* in elastomers, paints or cosmetics) as well as for understanding their bio-interactions. Here, the effect of synthesis conditions on the silica surface hydroxylation is investigated. Furthermore, a correlation between silanol content/type and *in vitro* cytolytic activity of flame-made silicas is sought by measuring the induced membrane damage in human blood monocytes. By controlling the flame synthesis variables, silica nanoparticles with the same specific surface area but distinct surface chemistry are prepared and characterized by Raman and infrared spectroscopy, thermogravimetric analysis, and titration with lithium alanate. The surface silanol content of silicas made in relatively short and "cold" flames is rather high (up to ~8 OH/nm^2) and is reduced down to ~4 OH/nm^2 for silicas made in long and "hot" flames. The latter is closer to that of commercial fumed silica. Moreover, "cold-flame" silicas exhibit higher proportion of hydrogen-bonded *vs.* isolated silanols and more physisorbed water than "hot-flame" ones of similar average primary particle diameter. Interestingly, the above surface chemistry differences translate to lower cytolytic activity for "cold-" than "hot-flame" silicas in human monocytes. Most notably, total silanol contents of up to ~17 OH/nm^2 are calculated by thermogravimetric analysis for "cold-flame" silicas, indicating their high internal silanol content. The latter provides new insight into the potential formation mechanism of silica particles in rapidly-quenched "cold" flames.

508192 Unraveling the Process of Ligand Adsorption to Heterogeneous Colloidal Substrates By Means of Catechols Binding to ZnO Nanoparticles (Invited)

Abstract:

Keywords: Functionalization, quantum dots, adsorption isotherm, binding enthalpy, ligand exchange reaction

Although being of major importance for various applications, the characterization of nanoparticle (NP) surfaces in dispersions, especially with respect to ligand exchange reaction, is still an open and highly challenging question. Therefore a general strategy to study the thermodynamics of ligand adsorption to colloidal surfaces was established by means of catechol (ethyl-3,4-dihydroxybenzoate, esterCAT) binding to ZnO NPs [1]. This combination represents a technically relevant model system as the catechol anchor group is a promising candidate anchor in applications like dye sensitized solar cells (DSSCs) [2].

First, isothermal titration calorimetry (ITC) was used to extract all relevant thermodynamic parameters, namely association constant, enthalpy, entropy and free energy of the ligand binding. To confirm the characterization of ligand binding by measuring the heat of adsorption, the free energy was cross-validated by mass-based adsorption isotherms. To close the mass balance, analytical ultracentrifugation (AUC) was applied to detect the amount of free, unbound catechol in solution. Then, Raman spectroscopy and nuclear magnetic resonance spectroscopy (^{1}H NMR) were performed to quantify the replaced amount of acetate which is the ligand from synthesis with esterCAT (65 %) and to distinguish bound (chemisorbed) and unbound (physisorbed) esterCAT. Finally, based on a collection of all our results, the in-depth picture of ligand binding to the esterCAT functionalized ZnO colloid was obtained.

In the second part of our work, this toolbox was applied to tailor the ZnO surface by CAT derivatives with different functionalities, namely hydrogen (pyroCAT), tert-butyl group (tertCAT), aromatic ring (naphCAT), ester group (esterCAT), and nitro group (nitroCAT). The results showed that the ZnO surface is highly heterogeneous and that the binding enthalpies of the CAT molecules follow the order of tertCAT < pyroCAT < naphCAT < esterCAT < nitroCAT, which is in agreement with the electronegativity of the tail groups. Moreover, the visible emission induced by defect sites on ZnO NPs surfaces was quenched in order of the binding strength of CAT molecules. Thus, by means of the toolbox, the correlations between electronegativity of molecules, binding enthalpies to the ZnO surface, and the photoluminescence (PL) properties of ZnO NPs were unraveled.

Finally, the kinetics of the ligand-induced defect quench were investigated in detail by two photon fluorescence (2-PF) spectroscopy, again using esterCAT as model functionality. It was found that at different ligand concentrations the fluorescence leveled off rapidly to a plateau within three seconds. Moreover, a full PL quench was observed at esterCAT concentrations as small as 0.25 monolayers (whereby 1 monolayer refers to the amount of molecules needed to cover the totally available ZnO surface area) provided, a concentration where maximum/equilibrium coverage of ZnO NPs with CAT molecules is clearly not yet reached. However, for liquid-borne ZnO NPs, there are various heterogeneities like different kinds of defects, surface occupied by solvent, as well as surface occupied by synthesis residuals or hydroxyl groups. Among all those, by 2-PF only the kinetics of esterCAT molecules binding to the defect sites were resolved whereby for Langmuir fitting it had to be assumed that the fluorescence intensity is proportional linearly with the number of defect sites. However, then the rate constant for defect saturation became accessible.

In conclusion, first, a widely-applicable strategy to study the thermodynamics of ligand adsorption to colloidal surfaces was established. Then, the toolbox was applied to tailor colloidal ZnO NP surfaces by CAT derivatives with different electronegativity. Finally, the electronegativity of ligands was linked with the binding enthalpy and PL quenching of NPs. We believe that such correlations are an important step towards a more general way of selecting and designing ligand molecules for surface functionalization. This allows establishing strategies for tailored colloidal NP surfaces beyond design on a case by case basis.

Acknowledgements

The authors want to thank the funding of Deutsche Forschungsgemeinschaft (DFG) through the Cluster of Excellence â €œEngineering of Advanced Materialsâ€ and project PE 427/29-1.

Literature

[1] W. Lin, J. Walter, A. Burger, H. Maid, A. Hirsch, W. Peukert, and D. Segets, Chem. Mater., 27 (2015) 358-369

[2] J.-F. Gnichwitz, R. Marczak, F. Werner, N. Lang, N. Jux, D.M. Guldi, W. Peukert, and A. Hirsch, J. Am. Chem. Soc., 132 (2010) 17910-17920

508491 Stealth Nanoparticles for Tumor Targeting: *in vivo* and *in vitro* Characterization (Invited)

Abstract:

Silica nanoparticles with proper surface modifications can sufficiently mimic chemical robot - an intelligent particle which can “find, visualize and cure” the tumor and thus might be used in the targeted drug delivery research. The successful tumor targeting is limited by not only the targeting technique and particle size but also the invisibility for the immune system ("stealth"), which goes hand in hand with the ability of nanoparticles to circulate in the bloodstream for more than few minutes. The presented work focuses on the preparation, characterization and *in vivo*testing of fluorescently labeled silica nanoparticles modified with stealth poly (N-(2-hydroxypropyl) methacrylamide) (pHPMA) and monoclonal antibody IgG M75 against Carbonic Anhydrase IX (CA IX, targeting to the hypoxic tumors like colorectal adenocarcinoma).

Fluorescently labeled (fluorescein isothiocyanate) silica nanoparticles with mean diameter of ≈170 nm and zeta potential between -15 and -20 mV exhibited great colloidal stability in buffers and physiological media when modified with pHPMA. The amount of pHPMA and IgG M75 on the nanoparticle surface was evaluated via amino acid analysis with resulting 3.6 w.% and 1.0 w.%, respectively. Further *in vitro* characterization employing flow cytometry and two different cell lines (HT-29: expresses CA IX; DLD-1: does not express CA IX; negative control) showed specific interaction of IgG M75-modified nanoparticles with HT 29 cell line while no nonspecific interactions were detected. Finally, the *in vivo* study on Nu-Nu nude mice with tumors grown from HT-29 cell line revealed that nanoparticles bearing IgG M75 antibody can accumulate in the tumor in tenfold higher concentration and are retained considerably longer than nanoparticles without specific targeting. Furthermore, the nanoparticles were detected in the blood even after 60 minutes after the intravenous administration, which might be contributed to the stealth pHPMA coverage, providing the sufficient contact with the organism and thus increases the possibility to reach the target.

508397 Heterogeneous Liposome Assemblies: When Soft Meets Hard (Invited)

Abstract:

Liposomes, lipid bilayer vesicles, are almost routinely used as encapsulating agents for drug delivery. Individual liposomes typically serve as a protection layer that prevents the encapsulated drug from premature release and protects it from the environment. The release of the drug is then relatively slow and spontaneous. Although intensive research is carried on in order to prepare systems that are sensitive to changes in pH, temperature or external stimulus such as sonication or magnetic field, there is still lot of unexplored possibilities.

In our work we have designed liposome-based entities that consist from one or more liposomes as compartments and can contain also iron oxide nanoparticles (MgNPs) for remote control of the aggregate and/or silica nanoparticles to improve the loading capacity. We have tested several different approaches of forming the liposomal aggregates, including use of the alginate microcapsules as liposomes â€“ MgNPs vehicle, charged liposome aggregation due to high ionic strength of the buffer, using DNA amphiphiles as anchors between liposomes, and using polyelectrolytes such as poly-L-lysine as a glue for charged liposomes and MgNPs. This last approach was studied extensively and resulting aggregates, stabilized using the MgNPs, were tested for their ability to follow the magnetic field and to perform remotely triggered release. We have also tested the method of increasing the loading capacity of individual liposomes using porous silica nanoparticles (due to their good sorption properties). Such approach can significantly increase the loading capacity of liposomes and influence their response to external stimulus.

508570 Magnetic Iron Oxide Nanoparticles As a Local Source of Heat for Remotely Controlled Reaction

Abstract:

Magnetic iron oxide nanoparticles as a local source of heat for remotely controlled reaction

Ayse Beyza Aysan[1], Zdeněk Knejzlík[1,2], Aleš Zadražil[1], František Štěpánek[1]

[1]University of Chemistry and Technology, Faculty of Chemical Engineering Department, Prague, Technicka 5, 166 28 Prague 6, Tel. +420 220444320, E-mail: aysana@vscht.cz;

[2]Institute of Organic Chemistry and Biochemistry of the Czech Academy of Sciences of the Czech Republic, Flemingovo náměstí 542/2, 166 10, Prague 6

Nanomaterials have seen considerable attention in biochemical processes due to their physical, chemical and biomedical properties. One of the promising classes of nanomaterials are magnetic nanoparticles with the ability to undergo down-stream modification making them biocompatible, nontoxic, high level accumulated in the target region and acting as a local source of heating in the presence of an radiofrequency alternating magnetic field (RFAMF). High surface area and engineered functional groups make from the nanoparticles an ideal tool for magnetic resonance imaging (MRI), drug/gene carrier, tissue repair or heating mediators in hyperthermia.

The aim of the presentation is to use the colloidal magnetic nanoparticle suspension to initiate the polymerase chain reaction (PCR). Firstly, magnetic iron oxide nanoparticles (MNPs) were fabricated with relevant specifications and combined with PCR mixture containing all required components. Concordantly, we present a research about compatibility of MNPs with the PCR components and define the conditions for an effective PCR in the presence of high concentration of MNPs and regulate temperature cycles for PCR by RFAMF.

MNPs were fabricated successfully with different types of functional groups including sodium citrate dehydrate ($Na_3C_6H_5O_7.2H_2O$), 3-aminopropyl-triethoxysilane (APTES) and dextran 70 (from Leuconostoc spp., approximately Mw 70 kDa). After characterization of MNPs in terms of concentration, size distribution, surface morphology, and radiofrequency heating ability, they were used in PCR reaction to see the effect on the process. According to the results the response to the RF alternating magnetic field is proportional to the size and concentration of the nanoparticles. Dextran coated MNPs were stable in the PCR reaction mixture even in high concentration while the other types formed aggregates in excess of 1 µm. Experiments were focused on dextran coated MNPs in sufficiently high concentrations to achieve the required PCR temperature by exposure to an RF alternating magnetic field. Preliminary results showed that dextran coated MNPs in 5mg/ml concentration could increase the reaction temperature up to 97°C which was sufficient to start the PCR reaction under RF alternating magnetic field.

508617 Multi-Scale Engineering of Biomedical Made Nanomaterials and Devices By Scalable Flame Synthesis

Abstract:

In the last decades, nanostructured materials have demonstrated a considerable potential in many medical applications. This ranges from targeted drug delivery[1] to non-invasive medical diagnostics[2]. For example, the synthesis of stimuli-responsive nanocomposite able to more easily penetrate cell membranes and release specific molecules[1] is of great significance for delivery of anti-cancer drugs. Similar achievements have been obtained for medical devices[2, 3]. Portable diagnostic methods have been empowered by the utilization of tailored nanostructures[4]. For example, the latter can amplify relatively week biomarkers such as breath and body odour analysis resulting in novel non-invasive medical diagnostics opportunities[5].

We have recently shown that a large variety of nanoparticles such as photo-responsive TiO_2 and chemo-responsive SnO_2 can be rapidly synthesized by combustion of organometallic precursor solutions. This flame synthesis process is a very scalable technology currently utilized for the production of most nanoparticle commodities such as Carbon Black, fumed Silica, and Titania pigments. Its application to biomedical engineering has led to the synthesis of tailored materials with unique chemical and physical properties such as surface enhanced SnO_2-TiO_2 solid solutions[7]. These flame-made nanomaterials have demonstrated excellent gas sensing properties toward a large number of reducing and oxidizing gases[8]. This resulted in the first portable breath analysers for acetone detection, the main breath-marker for diabetes[2]. Optimization of this synthesis approach has enabled the fabrication of nanostructured surfaces for numerous applications including batteries, solar cells, and super-hydrophilic / hydrophobic[9] coatings[10]. In particular, such three-dimensional nanostructured surface allows enhancement of substrate - surrounding interaction and thus have potential to improve interface properties at sub-cellular level.

Here, we will present the synthesis of tailored nanostructured material by flame synthesis. We will focus on the application of this scalable approach for three main different applications, namely, the fabrication of stimuli-responsive nano-substrates for drug deliver and imaging, the assembly of biosensors for non-invasive medical diagnostics, and the rapid fabrication of biocompatible coating for medical implants. We will discuss the multi-scale optimization of the material nano-micro structure demonstrating that the interaction between organic and inorganic components can be optimized through suitable hierarchical configurations. References

1. H. Lee, et al., *Mol Pharm*, 2010, **7**, 1195-1208.

2. M. Righettoni, et al., *Anal Chim Acta*, 2012, **738**, 69-75.

3. M. Righettoni, A. Tricoli and S. E. Pratsinis, *Chem. Mat.*, 2010, **22**, 3152-3157.

4. A. Tricoli, M. Graf and S. E. Pratsinis, *Adv. Funct. Mat.*, 2008, **18**, 1969-1976.

5. G. Peng, et al. *Nature Nanotech.*, 2009, **4**, 669-673.

6. A. Tricoli, M. Righettoni and S. E. Pratsinis, *Nanotechnology*, 2009, **20**, 315502.

7. A. Tricoli and S. E. Pratsinis, *Nature Nanotech.*, 2010, **5**, 54-60.

8. A. Tricoli, et al., *Nanotechnology*, 2010, **21**, 465604.

9. A. Tricoli and T. D. Elmoe, *AIChE Journal*, 2012, **58**, 3578-3588.

509642 Targeting and Killing Melanoma Cancer Using Photoactivated Nanoparticles

Abstract:

Photodynamic therapy (PDT) uses photoactivated nanoparticles that are functionalized to target cancer cells and applies light to excite the nanoparticles which subsequently release reactive oxygen species (ROS) that set off a process resulting in cell-suicide of the cancer cells only. This study tested the polymer Poly[2,1,3-benzothiadiazole-4,7-diyl[4,4-bis(2-ethylhexyl)-4H-cyclopenta[2,1-b:3,4-b']dithiophene-2,6-diyl]] (PCPDTBT) for use as the photoactivated nanoparticle while also examining the Endothelin 3 ligand (ET3) for use as the functionalizing agent for targeting melanoma. This study was inconclusive. Although PCPDTBT was likely an effective photosensitizer for PDT, the results suggested that the functionalized nanoparticles did not target A375 melanoma cancer cells, likely due to the aggregation of the nanoparticles which exceeded the largest size that can be transported into cells at sizes of 100 nm and larger.

508595 ALD-Formed Cobalt/Alumina Nanostructures Active for Fischer-Tropsch Synthesis

Abstract:

Cobalt metal was deposited onto alumina (Al_2O_3) supports using atomic layer deposition (ALD) to create dispersed catalysts for Fischer-Tropsch Synthesis (FTS). We show that the phase of the Al_2O_3 support greatly influences the FTS activity of the resulting catalyst. ALD literature predicts that Co deposited on Al_2O_3 with low cycle numbers should only produce Co nanoparticles less than 2 nm in diameter, which fall into the well-characterized non-active region of nanoparticle sizes for FTS. Our FTS experiments confirm that this inactivity is indeed the case for Co deposited on amorphous Al_2O_3. However, Co deposited on calcined Al_2O_3 is in fact active, suggesting an alternative ALD deposition regime not previously recognized. Aberration corrected STEM/EELS imaging shows that in addition to nanoparticle deposition, cobalt has also deposited on Al_2O_3 as semi-continuous planar layers likely less than 3 monolayers thick, which are likely responsible for the catalystâ€™s FTS activity. Further, it was found that the number of ALD cycles alters the activity of these catalysts, with 4 and 8 cycle catalysts being more active than the 1 cycle catalyst. These data suggest Co requires the presence of crystalline Al_2O_3 surfaces in order to achieve a thin semi-continuous planar growth regime. Finally, density functional theory (DFT) calculations were carried out to understand the thermodynamics of Co small particle vs. planar layer (monolayer and bilayer) growth on Al_2O_3 and explain the activity shown by these catalysts.

508358 Synthesis and Characterization of Polyester and Polyamide Microcapsules for Vitamin E Encapsulation

Abstract:

Vitamin E (α-tocopherol) is a strong antioxidant extensively used in numerous medical and cosmetic applications, e.g. in the prevention and treatment of Alzheimer's disease and other neurological diseases. Since vitamin E is considerably sensitive to light, heat and oxygen, and therefore suffer rapid degradation, its protection before use is crucial.

Microencapsulation is an appropriate solution for that protective function. A well-established strategy, microencapsulation used in several areas for the transport and protection of active agents. This approach allows the retention of the agent of interest, protecting it from degradation, evaporation and, more importantly, it offers the possibility of controlled release.

In this work, vitamin E was microencapsulated, by interfacial polycondensation and solvent diffusion method, to form solid resistant capsules, composed of polyester either polyamide walls. Different conditions were used to achieve the best microencapsulation formulation. The produced microcapsules were characterized in terms of size distribution, shape ad storage stability. The encapsulation efficiency and kinetics release of vitamin E were also evaluated.

Both polyester and polyamide based microcapsules showed to be spherical, with a smooth surface and unimodal size distributions. Additionally, satisfactory encapsulation efficiencies of Vitamin E and slow core release profiles were obtained.

The produced microcapsules demonstrated great potential as control delivery systems for pharmaceutic, textile, food or cosmetic applications.

References

J. Wei, Z. Li, L. Liu, and X. Liu, "Preparation and Characterization of Novel Polyamide Paraffin MEPCM by Interfacial Polymerization Technique," pp. 4588–4593, 2013.

K. Hirech, S. Payan, G. Carnelle, L. Brujes, and J. Legrand, "Microencapsulation of an insecticide by interfacial polymerisation," vol. 130, pp. 324–330, 2003.

D. Saihi, "Microencapsulation of ammonium phosphate with a polyurethane shell . Part II . Interfacial polymerization technique," vol. 66, pp. 1118–1125, 2006.

G. P. Yao, J. H. Zheng, and H. F. Chen, "Preparation and Application of VE Microcapsules with Polyurethane Shell for Skin-Care Textiles," *Adv. Mater. Res.*, vol. 331, pp. 402–406, 2011.

A. M. Samuni, A. Lipman, and Y. Barenholz, "Damage to liposomal lipids: protection by antioxidants and cholesterol-mediated dehydration.," *Chem. Phys. Lipids*, vol. 105, no. 2, pp. 121–134, Apr. 2000.

M. Stumpo *et al.*, "Scale-up of polyamide and polyester Parsol ® MCX nanocapsules by interfacial polycondensation and solvent diffusion method," *Int. J. Pharm.*, vol. 454, no. 2, pp. 678–685, 2013.

505783 Green Synthesis, Characterization and Physical Properties of Silver Nanoparticle Embedded in PVA

Abstract:

Silver nanoparticles (Ag NPs) were synthesized by green synthesis using *Malva sylvestris* extract as reduced and capping agent added to poly vinylalcohol (PVA) solution and monitoring the growth of Ag NPs after 2, 4 and 8 min where films were made by casting method as Ag NPs PVA composites. Brownish to radish films are prepared and characterized by SEM, EDS and UV vis spectra of absorbance and fluorescence. AC and DC conductivity and mechanical properties by ultrasonic waves were studied. SEM shows spherical shape of Ag NPs. According to UV vis spectra, peak of plasmon resonance among 350- 600 nm increased with time of synthesis, also fluorescence emission at 443 nm as an indication of successive synthesis of Ag NPs. AC and DC conductivity were studied and Ag NPs and increased with increasing of frequency and temperature respectively.

539154 Hydrogenolysis of 5-Hydroxymethylfurfural to 2,5-Dimethylfuran Over Supported Pt-Co Bimetallic Catalysts Under Mild Conditions

Abstract:

Highly dispersed Pt-Co bimetallic catalysts were deposited on multi-walled carbon nanotubes (MWCNTs) by atomic layer deposition. High-resolution TEM and TPR analysis verified the formation of Pt-Co bimetallic particles. Catalysts were applied for hydrogenolysis of 5-hydroxymethyfurfural (HMF) to 2,5-dimethyfuran (DMF). High yield of DMF (> 90%) was achieved in hydrogenolysis of HMF over the optimized Pt-Co/MWCNTs catalyst after 8 hr of reaction time under 10 bar H_2 at 160 °C. Through a series of experiments and comparison, the synergistic effect among Pt, Co, and MWCNTs was investigated. The results revealed that the synergistic effect between Pt-Co and MWCNTs played an important role in the improvement of selectivity to DMF for Pt-Co/MWCNTs bimetallic catalysts. In addition, steric hindrance appeared when the Co loading in Pt-Co/MWCNTs was high and it affected the activity of Pt-Co bimetallic catalysts. However, the moderate activity can inhibit the production of byprodct and thereby improve the yield of DMF.

524844 Atomic Layer Deposition (ALD) Coatings Enable Higher Energy Batteries with Enhanced Lifetimes

Abstract:

Lithium-ion batteries are driving todayâ€™s growth in the battery market. Dramatic changes and industry expansion are creating large new markets for cost-effective, nano-engineered materials. ALD Nano uses proprietary Atomic Layer Deposition (ALD) techniques to enable new battery materials that solve battery performance challenges cost-effectively at scale.

506273 Increasing Durability of Fuel Cell Catalysts with Atomic Layer Deposition

Abstract:

The proton exchange membrane (PEM) fuel cell is a highly efficient, portable, non-greenhouse gas emitting source of electrical power that holds great promise as a replacement for the internal combustion engine. One of the most significant challenges in the development of the PEM fuel cell is the long-term durability of the oxygen reduction reaction (ORR) catalyst. Cathode potential cyclingâ€"resulting from the variable voltage loads imposed during vehicular operationâ€"is known to promote agglomeration of the Pt nanoparticle catalyst and degradation of the carbon support. In this work, atomic layer deposition (ALD) was used to produce both modified Pt nanoparticle catalysts as well as PtNi extended surface catalysts with enhanced electrochemical durability. Commercial and ALD-synthesized (on functionalized carbon black) Pt/C catalysts were modified with TiO_2 ALD to create protective nanostructures on the support surface that acted as physical barriers to Pt agglomeration and preserved electrochemical surface area and mass activity over voltage cycling. Extended surface catalysts synthesized by galvanic displacement have previously been shown to have high activity and enhanced durability over Pt/C catalysts due to highly-coordinated surface Pt and elimination of the carbon support. An analog of the galvanic displacement process was developed using Pt ALD on Ni nanowires substrates to create a novel extended surface catalyst with mass activity almost 3X higher than benchmark Pt/C and over 70% retention of activity after durability testing. The PtNi nanowire catalysts were characterized at 5, 10, 20, and 30 cycles of ALD to analyze the nucleation and growth of the Pt during the deposition process and its effects on active surface area.

507254 Evaluating the Influence of Dry Granulation Processing Variables on the Rheological Properties of Granules

Abstract:

Granulation is often employed in the production of pharmaceutical formulations for oral solid dose delivery. It is most often required when constituent materials in a formulation have poor flow properties, enabling these components to be combined to improve process performance and ensure content uniformity.

Dry granulation, based around the use of a roller compactor and integral mill/screen, can have significant advantages over more common wet granulation techniques. Dry granulation can reduce processing costs and can also be used with moisture/solvent/heat sensitive materials. Unlike a wet granulate, the dry granulate does not require additional drying and milling before encapsulation or tabletting.

Dry granulation is not suitable for all types of material, however, there is little research into which properties make a formulation suitable for this process. In addition, there is little to indicate which process parameters produce optimal granulate quality in order to achieve interruption free processing and high quality final products.

In this study an FT4 Powder Rheometer® (Freeman Technology, Gloucestershire, UK) was used to demonstrate how processing parameters influence the properties of a placebo formulation granulate. The granulate was produced using a roller compactor (MiniPactor, Gerteis Maschinen+Processengineering AG, Switzerland) on which the roll gap and compaction force were varied. The dynamic flow, bulk and shear properties of the resulting granulates were then evaluated using the FT4 Powder Rheometer.

The data clearly demonstrates that several rheological properties, known to influence die filling efficiency and Critical Quality Attributes (CQAs) of the resulting tablets, could be controlled by changes in roller compactor settings.

This study emphasises the importance of assessing powder properties using techniques that closely simulate process conditions and how it is possible to identify Critical Process Parameters to optimise a dry granulation process and ensure high final product quality.

507489 Assessment of the Homogeneity of an Alumina Powders Mixture Using Rheological Parameters

Abstract:

This study has been carried out in order to improve the comprehension of powder mixing in the nuclear fuel manufacturing process. Thus, a binary mixture of two Alumina (Al_2O_3) powders, GE15 and CR6, was used to simulate the oxides powders actually involved in the process. Different mixing technics and parameters were tried and compared in terms of homogeneity of the mixtures produced. The concept of intensity of segregation [1] was used for assessing mixture homogeneity through a sampling protocol and classical statistical analysis tools [2]. One of the difficulties of working with these compounds is that, due to the similarities between CR6 and GE15 powders (same chemical composition, same colour, etc.), the direct measure of the composition of a given sample is not possible. Since homogeneity of a mixture is assessed from the variation of the composition between samples, it is important to find an indirect way to measure the composition of an Alumina powders mixture. Therefore, different correlations between measurable parameters (specific surface area, Sauter diameter, cohesion and flow function coefficient) and the composition of the mixtures were found empirically allowing indirect measurement of the composition of a given sample.

The four studied parameters: specific surface area (, measured by nitrogen adsorption using BET method), Sauter diameter (, measured by laser granulometry, dry transportation with air at 0.1 bar), cohesion and flow function coefficient (and , measured by FT4 shear test at 9 kPa on a 10 mL cell), were compared in terms of precision, relevance and feasibility.

It appears that the homogeneity assessment methods involving rheological parameters (cohesion and flow function coefficient) show many advantages compared to the other methods. Not only are they the most accurate correlations found but both parameters are measured simultaneously, meaning that it is possible to apply both methods on the same samples while all the tests (N_2 adsorption, laser granulometry and FT4) are destructive. Those two methods also allow the estimation of the homogeneity at a more interesting scale (around 7 g of powder), indeed, specific surface area and Sauter diameter methods give a homogeneity at a scale of less than one gram which is not relevant for the end-use properties of the final product.

[1] P. Danckwerts, The definition, measurement and some characteristics of mixtures, Applied scientific research. Section A, 3 (1952) 279-296

[2] S. Massol-Chaudeur, H. Berthiaux, S. Muerza, et J. Dodds, A numerical model to identify the structure of a high-dilution powder mixture, *Powder Technol.*, 128 (2002) 131-138

508516 Characterising the Grain Shape: In Search of Size Independent Shape Descriptors

Abstract:

Granular materials contain a large number of grains that differ in size and shape. While the characterisation of the size distribution is typically done using sieve analysis, the characterisation of the shape of the grains remains a difficult task and is one of the major bottlenecks in the characterisation of the granular material. In this study, we attempt to characterise the shape of three different types of grains commonly used in iron and steel making industry, namely pellet, sinter and iron ore lump. While the pellet particles can be considered nearly spherical in shape, imposing the spherical shape on the iron ore lumps and highly irregular sinter particles seems a very poor choice. We consider two different size ranges of pellet particles and four different size ranges for sinter and iron ore lumps. Imaging of the grains is done using an iPhone 7 Plus camera and sufficient number of grains of each type in all the size ranges are considered. We perform the image analysis and obtain various boundary parameters (such as projected area, minor and major axis, perimeter, etc.) of the grains using MATLAB image processing toolbox. In addition, we define two additional boundary parameters to characterise the deviation of the projected area from that of a circle and calculate their values for all the different type of grains considered in this study. Using these length scales, we define a total of nine different shape descriptors and obtain the distribution of these shape descriptors for each type of grain for all the size ranges. Interestingly, for a given type of grains, the frequency distribution curves for these nine shape descriptors turn out to be nearly independent of the size range of the grains. This suggests that the chosen shape descriptors indeed represent properties related to the shape of the grains alone, irrespective of the size of the grains considered.

We investigate the correlation among these shape descriptors by evaluating the values of the correlation coefficients between each pair of the shape descriptors using the Pearson method, Kendall method and Spearman method. Our results indicate that out of the nine shape descriptors considered in this study, five shape descriptors are strongly correlated with each other and only the remaining four shape descriptors need to be considered in characterising the shape of the grains. Out of these four uncorrelated shape descriptors, two shape descriptors, namely the aspect ratio of the grains and the root of form factor, have been considered important by researchers in previous studies. The two other shape descriptors proposed in this study, which are used to characterise the deviation of the projected grain shape from that of a circle, are not measured previously to the best of our knowledge. We compare these four shape descriptors for the pellets, sinter and iron ore lumps having the same sieve size range and show that the shape descriptors for these three different types of grains are different.

We next calculate the values of these four shape descriptors for relatively larger coke particles using 2d image analysis. We also perform 3d image analysis of some of these coke particles and calculate the values of the analogous shape descriptors in 3 dimensions. We investigate if the identified 2d shape descriptors (obtained from 2d imaging) can be correlated to the analogous 3d shape descriptors (obtained from 3d imaging) and explore if it is possible to use the 2d imaging technique to estimate the 3d shape descriptors, thus eliminating the need of 3d imaging.

508543 Investigating Mechanical Properties of Surfactant Films at the Solid-Liquid Interface Using AFM

Abstract:

Corrosion is a well-known problem in the oil and gas sector and also in many other industries. Surfactant based inhibitors have been used for corrosion control in oil pipelines but their inhibition mechanism is still not fully known. Our group is studying surfactant corrosion inhibition mechanism by various characterization techniques (AFM, Electrochemical and Spectroscopic techniques). A better understanding of corrosion inhibition mechanism will lead to the design and development of not only cost-effective but also greener surfactants system for corrosion inhibition purposes.

Atomic force microscopy (AFM) is an effective tool to study the barrier properties of surfactant films at the solid-liquid interface. Determination of mechanical and topographical properties of single and mixed surfactant coatings by atomic force microscopy (AFM) can delineate the role of physical vs. electrochemical barrier by such films in the corrosion inhibition process. . So far, the AFM data, obtained using a closed cell, indicate the effect of surfactant concentration and that of hydrocarbon chains interactions on the mechanical properties of cationic surfactant films. Ultimately, integration of the AFM and electrochemical results will reveal the correlation between the surfactant structures at the solid-liquid interface and their corrosion inhibition efficacy.

508545 Deposition Rate Consequences of the Formation of Multi-Spherule Cluster Aggregates in Gases "Role of Momentum Shielding™"

Abstract:

For engineering flow systems characterized by particulate suspensions in gases, will the formation of *cluster aggregates* (CAs) in the mainstream accelerate deposition rates on immersed targets or confining walls ---or is aggregate formation more likely to suppress capture rates? Recent research on the drag properties of large cluster aggregates (e.g., stimulated in part by the Sorensen (2011) review) has opened the door to answering these important mechanism- (and pressure-) dependent questions. Sub-100 nm "primary" spherules, which may find themselves in even tenuous 'fractal-like' N-spherule clusters, can take advantage of what we call "momentum shielding"---analogous to the drag advantages experienced by birds who elect to fly in formation. We show here that the extent of this "momentum shielding" can be conveniently quantified (in a dimensionless function we call: $S_{mom}(N;Kn_1;\text{aggregate structure}))$, and used to predict the deposition-rate consequences of aggregation in aerosol flow systems when the cluster deposition mechanism is dominated by either: i) isothermal convective diffusion (C-D), ii) thermo-phoresis (TP-) or iii) inertial impaction (II). As expected, we show that for any type of aggregate structure (including 'fractal-like'), momentum shielding increases with cluster size (e.g. N) and gas pressure (via the reciprocal Knudsen number based on spherule radius R_1). We also show how S_{mom} can be calculated for the limiting case of a N>>1 random close-packed-D_f=3 cluster ---which is nearly gas impermeable. For fractal-like aggregates, including DLCAs (D_f=1.8) and RLCAs (D_f=2.1) we have recently shown that the abovementioned function: $S_{mom}(N, Kn_1; D_f)$ is actually sufficient to make useful engineering estimates of CA Brownian diffusion coefficients (Rosner and Tandon (2017b), thermophoretic diffusivities (Rosner and Tandon (2017a) and aggregate "stopping times" (Rosner and Tandon (2017c)----each in terms of their previously known values (D_1, $\alpha_T D_1$, and $t_{p,1}$) for isolated individual spherules in the same environment.

This information now enables instructive engineering calculations of the deposition-rate consequences of spherule aggregation to form large cluster aggregates-----i.e. at the same mainstream spherule volume fraction we can now determine how much aggregation will increase or decrease the total spherule acquisition rate on an immersed target. We present here representative results for each of the abovementioned particle transport mechanisms----i.e.: i) isothermal convective-diffusion (C-D), ii) thermo-phoresis (TP-) or iii) inertial impaction (II)----- including a brief account of the consequences of dealing with Kn1-dependent coagulation-'aged' aggregate size distributions (ASDs) in the mainsteam. We summarize our current re-examination of the canonical problem of inertial impaction on a circular cylinder in Re>>1 crossflow (Rosner and Tandon (2017c), dealing now with CA-"projectiles" but in the absence of spherule loss from the much larger target. We conclude by briefly outlining our planned work on the efficient "inversion" of sampled cluster-ASD-data to reconstruct, using the abovementioned $S_{mom}(N,Kn_1;D_f$ information, what ASD must have existed in the mainstream---at least in the absence of appreciable supercritical aggregate break-up.

References:

1. Rosner, D. E. and P. Tandon (2017a). Knudsen Transition Effects on the Thermophoretic Properties of Fractal-like Aggregates; Implications for Thermophoretic Sampling of High-Pressure Flames, Aerosol Science and Technology (in press).

2. Rosner, D. E. and P. Tandon (2017b). Aggregation- and Knudsen Transition Rarefaction- Effects on Particle Mass Deposition Rates by Convective-Diffusion, Thermophoresis or Inertial Impaction; Consequences of Multi-spherule 'Momentum Shielding', Aerosol Science and Technology (in press).

3. Rosner, D. E. and P. Tandon (2017b). Mainstream Aggregation and Polydispersity Effects for Inertially-Dominated Deposition Rates on Cylinders (or Leading 'Edges') in Crossflow"---- Consequences of 'Momentum Shielding' for the Impaction of Large, "Compact" ($2<D_f<3$) Multi-spherule Aggregates, Aerosol Science and Technology (prepared for submission, August 2017).

4. Sorensen, C. M. (2011). The Mobility of Fractal Aggregates: A Review, Aerosol Science and Technology. 45: 765-779.

508609 Experimental and Numerical Studies on the Thermal Diffusivity of Packed Powder Beds

Abstract:

Several industrial processes, such as calcination and drying, involve thermal treatment of particles. In order to achieve efficient processes and good quality products, the thermal properties of the particles and that of the bulk powder must be well defined. Previously, some models, such as the Krischer and Maxwell-Eucken models, have been proposed to relate particle thermal properties to that of the bulk powder. Though these models have been shown to work well, it is not very clear how to account for changes in the packing density, moisture content, and arrangement of particles. In addition, the particle thermal properties of most materials are not readily available, which limits the application of these models. To get a better understanding of the relationship between the particle thermal properties and that of the bulk powder and also to quantify the effects of packing density and moisture content on the bulk powder thermal properties, we run several experiments and discrete element method (DEM) based simulations. In both the experiments and simulations, a pipe is filled with particles where the temperature of the pipe is kept constant. We measured the temperature change in the powder bed and compute the actual thermal diffusivity of the powder bed. We tested several types of powders and glass beads in the experiments. The packing fraction, the moisture content, and size of particles were varied in the experiments. In the simulations, in addition to the packing fraction and particle size, the thermal properties of the particles were also varied. In this talk, we will present the details of the experimental and simulation results.

509556 Determination of Size, Size Distribution and Refractive Index of Artificial and Biological Microparticles

Abstract:

For the approval of clinical studies, it is mandatory to validate the production process of artificial oxygen carriers by reliably specifying the properties of the microparticles used like the newly developed blood substitute HbMP-700 (1-3). In particular, besides shape, size and size distribution, the amount of hemoglobin (Hb) and methemoglobin (metHb) content is essential.

We present a method based on the simultaneous measurement of collimated and diffuse transmission spectra in the visible and ultraviolet range and numerical analysis to infer the wavelength-dependent refractive index, size and size distribution of microparticles.

For the validation of the method, we used aqueous suspensions of monodisperse polystyrene microspheres of known concentration, determined by flow cytometric measurements. The related data analysis is carried out for synthetic and experimental data. For homogeneous spheres, Mie theory is applied to compute the solution for the scattering of a plane electromagnetic wave by a particle and to calculate the extinction cross section C_{ext}, i.e. the shadow cast by the particle as a function of the vacuum wavelength λ, the particle size, and the complex refractive index $n(\lambda)$. For dilute suspensions of particles, the measured quantity is an ensemble-averaged extinction cross section C_{ext} . This average is obtained by integrating over the particles' size distribution, which may not be known a priori. From the resulting spectral extinction cross sections, the complex refractive index and the mean and standard deviation of the size distribution can be inferred by means of nonlinear least-squares optimization. In order to reduce the number of free parameters of the problem to a feasible amount, $n(\lambda)$ is represented in a set of harmonic-oscillator-like resonance functions. The refractive index of polystyrene microspheres obtained is in good agreement with literature values for the bulk material, the mean and standard deviation of the distribution of particle *size D* are determined with an uncertainty in the range of a few nanometers. For example, for microspheres of mean diameter mean(D) = 2μm and distribution width std(D) = 9.4nm the uncertainty of the mean is $u\{\text{mean}(D)\}$ = 1.2nm and the uncertainty of the distribution width amounts to $u\{\text{std}(D)\}$ = 0.4nm.

In a second step, we applied the method to characterize HbMP-700 microparticles, approximately featuring a prolate spheroidal shape with 700nm length and 400nm width. Besides shape, size and size distribution, the amount of hemoglobin (Hb) and methemoglobin (metHb) content is essential. The relative concentrations of oxyHb, deoxyHb, metHb are accessible from spectral extinction and diffuse transmission measurements, since the different Hb variants exhibit different wavelength-dependent refractive indices. The method to infer size and refractive index described above can be applied to these non-spherical particles, since the deviation of their light scattering properties from those of spheres turns out to be small. The results based on the optical method will be compared to alternative procedures for HbMP particle characterization with respect to their size and size distribution, i.e. electron microscopy, centrifugation and dynamic light scattering. In addition, x-ray spectrometry will be used to detect the ratio between Fe^{2+} (oxy/dexyHb) and Fe^{3+} (metHb) as well as the iron concentration.

References

1) Novel hemoglobin particles--promising new-generation hemoglobin-based oxygen carriers. Bäumler H, Xiong Y, Liu ZZ, Patzak A, Georgieva R. Artif Organs. 2014 Aug;38(8):708-14. doi: 10.1111/aor.12331.

2) Nonvasoconstrictive hemoglobin particles as oxygen carriers. Xiong Y, Liu ZZ, Georgieva R, Smuda K, Steffen A, Sendeski M, Voigt A, Patzak A, Bäumler H. ACS Nano. 2013 Sep 24;7(9):7454-61. doi: 10.1021/nn402073n.

3) Hemoglobin-based oxygen carrier microparticles: synthesis, properties, and in vitro and in vivo investigations. Xiong Y, Steffen A, Andreas K, Müller S, Sternberg N, Georgieva R, Bäumler H. Biomacromolecules. 2012 Oct 8;13(10):3292-300. doi: 10.1021/bm301085x.

517131 FlowCam Nano ® Provides Counts, Sizes and Images of Nano-and Microparticles: Application to a Therapeutic Protein Pumping Study

Abstract:

Sub-visible particle characterization is becoming a more important method in assessing drug substance and drug product stability. However, the current USP protocol does not require any characterization of particles below 10 Î¼m. This greatly hampers our ability to properly assess protein stability as a function of particulate formation and obscures most sub-visible particles and various morphological properties.

In this study, we assess the capabilities of the new FlowCam Nano from Fluid Imaging Technologies, in conjunction with a pumping study that focuses on particulate formation generated during fill operations. Initial experiments show that the new FlowCam Nano can generate highly resolved images of nanoparticulates, allowing the simultaneous assessment of particulate morphology for nano- and microparticulates. This is extremely useful, as currently unpublished data (for the pumping study) show that as a function of post-pumping agitation, a significant increase in microparticle concentration coincides with a significant depletion of the nanoparticle concentration. This suggests that agglomeration (and probably not nucleation) is the more likely mechanism for larger microparticulate formation.

The ability to characterize the nano- and microparticulate morphology simultaneously, will allow us to better assess the cause of particulate formation (such as observing the transition of nanoparticles to microparticles and seeing if specific morphological properties of larger nanoparticulates are conserved in microparticulates). In addition, the FlowCam Nano is also sensitive enough to assess differences in the colloidal stability of a drug product in different formulations. During filling operations, and accelerated degradation studies, FlowCam Nano showed the differences in size distributions of IVIG in PBS and 0.25 M glycine buffer. Since IVIG is less colloidally stable in PBS, the size distribution varies significantly more per size bin and the overall mean effective spherical diameter is also ~ 600 percent greater after 24 hours of agitation in comparison to the size distribution data of IVIG in glycine. Thus, Flow Cam Nano has capabilities that are very useful in assessing drugs in the substance (e.g. optimizing drug formulations) and product (e.g. assessing particulates post fill & finish operations) phases.

Numerical and experimental estimation on the normal and tangential capillary bridge force adhered to two spheres

Kazuo Murase*, Keisuke Arai, Takato Ootsuka, Daiki Sakamoto and Futa Egawa,
Department of Applied Chemistry, Chuo University
1-13-27, Kasuga, Bunkyo-ku, Tokyo,1128551, Japan
**hmurase@kc.chuo-u.ac.jp*

Introduction

Liquid bridges extensively occur as a binder liquid adhered to solid surfaces in a variety of engineering applications such as crystal growth, ink-jet printing, material processing and powder technology. Especially the strength of adhesion and cohesion of liquid bridges adhered to particles contribute to the consolidation of an agglomerate in wet granular materials. Estimation of the attractive or repulsive force between particles is important in controlling kinetics of agglomeration and handling of granular materials. Characteristics of a liquid bridge e.g. the equilibrium bridge geometry [1], stability [2], static and dynamic binding force [3-6], rupture energy [7,8], and wetting hysteresis [9] etc., have been experimentally and numerically investigated.

The liquid distributions within a particles bed are classified into four types of categories, i.e. pendular, funicular, capillary and slurry states, according to the liquid contents [10]. The static normal capillary force at pendular liquid state have elaborately been investigated as Fisher [11], toroidal model [8] and the exact solution of Yung-Laplace equation [12]. The dynamic normal capillary forces [13] are experimentally estimated with lubrication theory. The tangential force of a liquid bridge [14] also contributes to the primary factor for the granular consolidation with the shear stress of wet granular materials but have not yet substantially been investigated.

This work aimed to estimate experimentally and numerically the normal and tangential capillary force of a static and dynamic liquid bridge adhered to two spheres. The asymmetry static liquid bridge with the shear stress demand for the exact solution of Young-Laplace equation for the equilibrium liquid drop including the effect of gravity.

1. Theory

1.1 Formulation of a asymmetry static liquid bridge

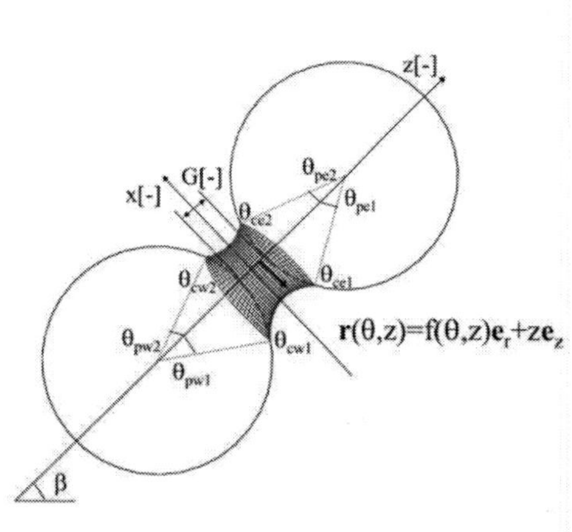

Figure 1 Geometry of a liquid bridge adhered to two spheres with arbitrary inclination in cylindrical coordinate system.

The asymmetry static bridge shape demands for the exact solution of the Young-Laplace equation due to the limit of the toroidal approximation. A prevalent geometry of a liquid bridge adhered to two spheres is drawn in Figure 1. The dimensionless position vector of a liquid surface in cylindrical coordinates is described as $\mathbf{r}(\theta,z) = f(\theta,z)\mathbf{e}_r + z\mathbf{e}_z$ divided by spherical radius, R', which $\mathbf{e}_r$, $\mathbf{e}_z$ are unit basis vectors in r-,z-directions. The symbol (') means the dimensional value. Fig.1 indicates the two dimensional front view with four half-filling (θ_{Pw1}, θ_{Pw2}, θ_{Pe1}, θ_{Pe2}) and contact angles (θ_{Cw1}, θ_{Cw2}, θ_{Ce1}, θ_{Ce2}).

The unit outward normal vector of a bridge surface, $\mathbf{n}_L$, is presented as follows [15],

$$\mathbf{n}_L \equiv \frac{r_\theta \times r_z}{|r_\theta \times r_z|} = \frac{f\mathbf{e}_r - f_\theta \mathbf{e}_\theta - f f_z \mathbf{e}_z}{[f^2(f^2+f_z^2)+f_\theta^2]^{1/2}} \quad \ldots(1)$$

,where low suffixes (r,θ,z) describe partial derivatives.

The dimensionless effective potential energy, $E=E'/\sigma R'^2$ with Lagrangian multiplier, $K=K'/\sigma R'^2$, on the constrained condition of a constant volume, $V=3V'/4\pi R'^3$, is described as follows [16],

$$E = \int_{-\pi/2}^{\pi/2}\int_{z_{min}}^{z_{max}} 2\left[f^2\left(f^2+f_z^2\right)+f_\theta^2\right]^{1/2} d\theta dz + \frac{Bo}{2}\int_{-\pi/2}^{\pi/2}\int_{z_{min}}^{z_{max}} (z\sin\beta + x\cos\beta) f^2 d\theta dz + K\left[V - \frac{1}{2}\int_{-\pi/2}^{\pi/2}\int_{z_{cp1}}^{z_{cp4}} f^2 d\theta dz\right] \quad \ldots(2)$$

Bond number (Bo) is a dimensionless number defined by $(\rho_L-\rho_G)gR'^2/\sigma$, where is the ratio of gravitational force to surface tension force equally as Etovoes number (Eoe).

The integral domain of a bridge surface is determined by the trapezoidal section of the upper and lower surface lines, [$z_{w2}<z<z_{e2}$ at $\theta=\pi/2$, $z_{w1}<z<z_{e1}$ at $-\pi/2$] and contact lines, [$z_{w1}<z<z_{w2}$, $z_{e1}<z<z_{e2}$ at $-\pi/2<\theta<\pi/2$] with four contact points settled by four half-filling angles ($\theta_{Pe1},\theta_{Pe2},\theta_{Pw1},\theta_{Pw2}$) in Fig. 1. Two contact lines on the upper and lower spheres were approximated by ellipse lines with parametric aspect ratios, which pass through respectively pairs of two contact points settled with half filling angles, ($\theta_{Pw1},\theta_{Pw2}$) and ($\theta_{Pe1},\theta_{Pe2}$). Filling angles, ($\theta_{PW}(\theta)$, $\theta_{PE}(\theta)$) on the lower and upper spheres have variable properties along the circumferential direction. The formulation of a static bridge shape at equilibrium state results in the extremum problem, which corresponds to the condition that the first order variation equals to zero, $\delta E=0$.

1.2 Estimation of the capillary force of a liquid bridge

The dimensionless mean curvature (H) and contact angles (θ_c) of a bridge surface are respectively calculated by eqs. (3),(4) once the outline of a liquid bridge is numerically evaluated.

$$-2H = \nabla_s \cdot \mathbf{n}_L = \left[(\alpha^2 + f_\theta^{\,2}) - f(f^2 + f_\theta^{\,2})f_{zz} - f(1+f_z^{\,2})f_{\theta\theta} + 2ff_zf_\theta f_{\theta z}\right]/\alpha^3 \quad \ldots(3)$$

$$,where\ is\ \alpha^2 = f^2(1+f_z^2) + f_\theta^2$$

$$\theta_c(\theta) = \cos^{-1}(\mathbf{n}_s \cdot \mathbf{n}_L) \quad \ldots(4)$$

,where contact angles also depend on the circumferential angle.

According to the Young-Laplace equation, the dimensionless pressure difference, $\Delta P= \Delta P'R'/\sigma$, across a liquid surface is related with the mean curvature of the surface.

$$\Delta P = (-2H) \quad \ldots(5)$$

The capillary force on a liquid bridge adhered to two spheres must be evaluated only on the contact line as known by the boundary method due to unspecified neck radius of a bridge and asymmetric condition. Dimensionless averaged capillary forces, $F_{Ci} = F'_{Ci}/(2\pi R'\sigma)$ on the upper and lower sphere (i=1,2 respectively correspond to lower, upper sphere) are presented as follows by three terms, which correspond to the surface tension, hydrostatic pressure and buoyancy terms along the longitudinal angle (θ). The sign of the capillary force set to be positive for attractive direction.

$$F_{ci} = \frac{1}{\pi}\int_{-\pi/2}^{\pi/2}\sin(\theta_{Pi})\sin(\theta_{Pi}+\theta_{Ci})d\theta - \frac{1}{2\pi}\int_{-\pi/2}^{\pi/2}\sin^2(\theta_{Pi})\Delta P d\theta \mp \frac{2}{3}\mathrm{Bo}\cdot V_{bi} \quad \ldots(6)$$

, which V_b is sphere volume submerged in a bridge

Finite Element Method is applied for the solution of the $\delta E=0$ equation. Isoparametric elements of a biquadratic polynomial shape function were selected for the integral domain and unknown radius. The set of non-linear algebraic equations for unknown radius at element nodes and Lagrangian multiplier was resolved by the Newton-Raphson method [9].

2. Experiment

Tensile strength of a static and a dynamic pendular liquid bridge

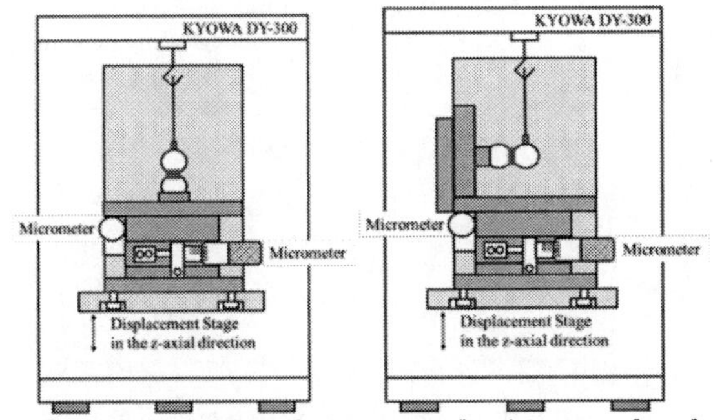

Figure 2 Experimental apparatus for (a) normal and (b) tangential tensile strength of a liquid bridge adhered to two spheres

Figure 2 shows the surface tensiometer (Kyowa interface Science Co., Ltd. DY-300) for measuring (a) the normal and (b) the tangential tensile strength of a water liquid bridge adhered to two stainless steel spheres (radius 3.0mm). The rod attached with one sphere is hanged up to the tensiometer and the other sphere fixed on the displacement stage is moved upward and downward by the tensiometer. The discrete and continuous displacement of the stage are performed for the normal and tangential deformation of a liquid bridge for the static and dynamic states. Two micrometers adjust the appropriate position of the lower spheres for the upper sphere. Water droplets are poured into the spheres aperture and substantially wetted by the displacement of the stage. The tensile strength at static and dynamic state and snapshots of liquid droplets by the digital camera and movie are recorded. The outlines of droplets are evaluated by a digitizer and estimate the half filling angles ($\theta_{Pe1}, \theta_{Pe2}, \theta_{Pw1}, \theta_{Pw2}$) and contact angles ($\theta_{Cw1}, \theta_{Cw2}, \theta_{Ce1}, \theta_{Ce2}$) at four contact points. The sphere configuration (radius, gap), the droplet volume, the half filling angles and the contact lines on spheres initiate parameters of numerical simulation.

3. Results and discussions

3.1 The normal tensile strength of a liquid bridge adhered to two vertical spheres.

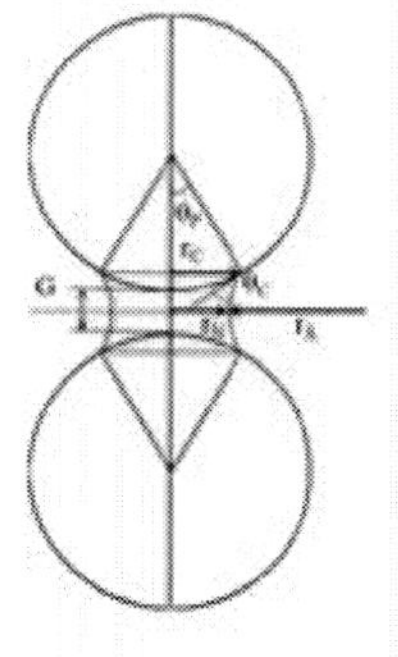

Figure 3 Geometry of toroidal model

The symmetric geometry of a liquid bridge adhered to two vertical spheres without gravity has been extensively investigated by toroidal approximation in Figure 3. The boundary method overestimates the capillary for of a liquid bridge rather than the gorge method but is feasibly and distinctly defined for the asymmetric bridge outline. Symmetric condition simplifies eq.(3) for the mean curvature and eq.(6) for the capillary force into eq.(7),(8) [7,16,17].

$$-2H = \nabla_s \cdot \mathbf{n}_L = \frac{1}{\alpha}\left(\frac{1}{f} - \frac{f_{zz}}{\alpha^2}\right), \textit{where} \quad \alpha^2 = 1 + f_z^2 \qquad \ldots(7)$$

$$F_{Boundary} = r_C \sin(\theta_{Pi} + \theta_G) - \frac{r_C^{\;2}}{2}(\Delta P) \mp \frac{2Bo}{3} V_{bi}, \quad r_C = \sin\theta_P \qquad \ldots(8a)$$

$$F_{Borge} = r_N - \frac{r_N^{\;2}}{2} \sin\theta_{cpi}(\Delta P) \mp \frac{2Bo}{3} V_{bi} \qquad \ldots(8b)$$

,where $r_N - r_A = \dfrac{A}{\tan(\pi/2 - \theta_p - \theta_C)} + \sin\theta_P, r_A = \left[A^2 + \{\sin\theta_P - (r_N - r_A)^2\right]^{1/2} A = G/2 + 1 - \cos\theta_P$

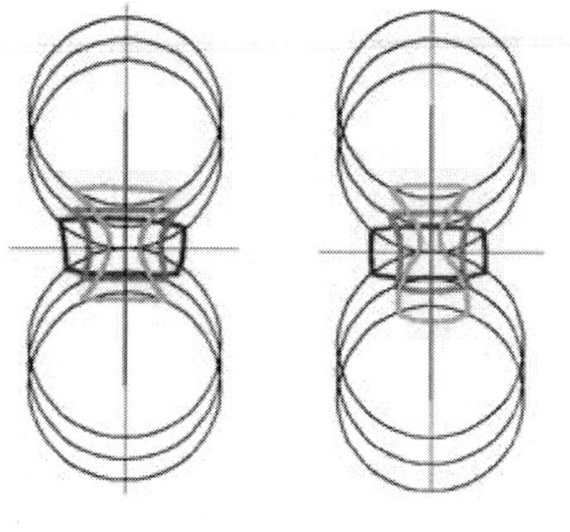

(a) Static state
G=0, 0.47, 0.91

(b) Dynamic state
G=0, 0.59, 1.16

Figure 4 Experimental droplet outlines at static and dynamic state, Bo=1.21,V=0.088

Figures 4 show symmetric outlines of liquid bridges around the vertical axis respectively at the static and dynamic state. Static state results in the shorter rupture gap and the larger dimensionless normal tensile strength from Figures 5a. Sphere gaps at the peak static capillary force depend on liquid volume. The contact angle on the upper sphere corresponds to the advancing contact angle at the higher Bo number due to gravity effect.

Fig. 4b indicates the downward transition of a liquid immediately before the rupture due to the wettability a liquid. The toroidal approximations over-estimate experimental data at the lower sphere gap but the sphere gaps at the peak dynamic capillary force depend on the velocity of stage displacement. The velocity reduces the peak value of dynamic capillary due to the downward transition of a liquid in Fig. 5b. The stepwise profiles of capillary force are due to the movement sensitivity of tensiometer for the stage displacement.

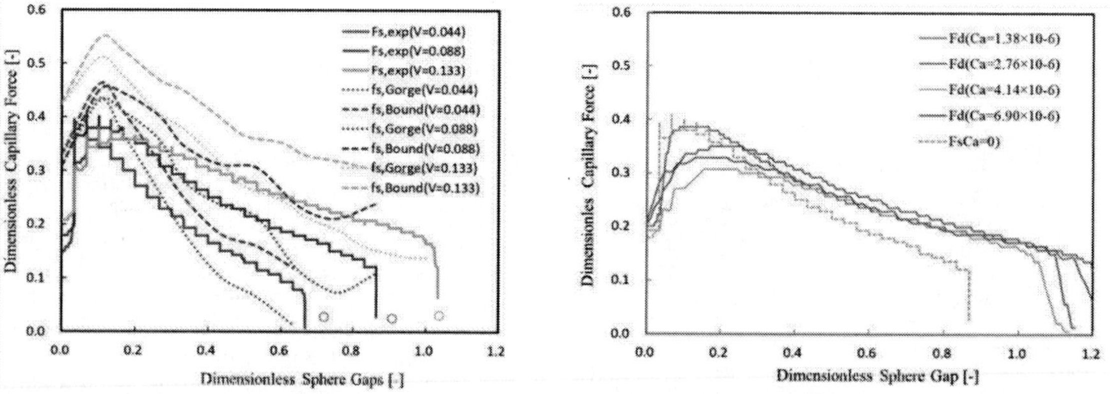

(a) at the static state (b) at the dynamic state

Figure 5 Effect of sphere gap on dimensionless capillary force

3.2 Estimation of boundary condition at contact line by LBM

The Young's equation describes the force balance among three interface tensions (solid-liquid, solid-gas and liquid gas) at three phase contact point and also defines the relationship between interface tension and contact angle. The solution of the Young-Laplace equation at the static state demands for one unknown boundary condition of either a contact angle or a contact line. The simulation for the wettability on three dimensional contact line demands for the elaborate process. Our experiment determines the contact lines on spheres under the constant liquid volume and is utilized for the boundary condition. Besides the perfect wettability the smooth transition of three phase contact lines is unexpected due to the advancing and receding contact angle. Lattice Boltzmann Method (LBM) is expected to estimate the boundary condition at static and dynamic state. The geometry of a liquid bridge is numerically estimated by FEM with half-filling angles predicted by 2 dimensional Shan & Chen model of LBM, which is used by D2Q9 model for transition of distribution function. The estimation for the liquid geometry by FEM has been verified. Figures 6 show the predicted liquid outlines by LBM and FEM. at a given liquid volume.

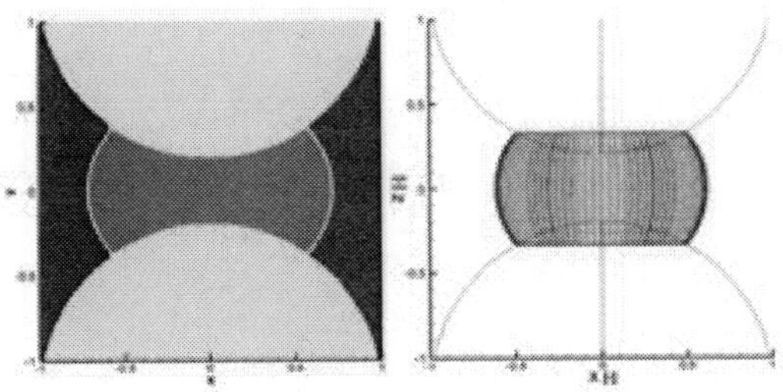

(a) Repulsive interparticle force

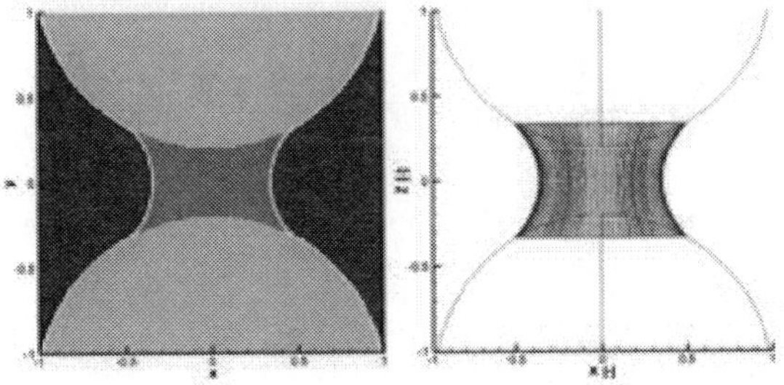

(b) Attractive interparticle force

Figure 6 Snapshots of liquid geometries at static state by LBM and FEM

Figures 7 show snapshots of dynamic liquid outlines at a given liquid volume up to the rapture state by LBM. The transition of boundary conditions, i.e. filling and contact angles, is smoothly described. Gravity and imperfect wettability, i.e. advancing and receding contact angles, are excluded in Figs. 7 but the effect of gravity on the liquid outline is avairable.

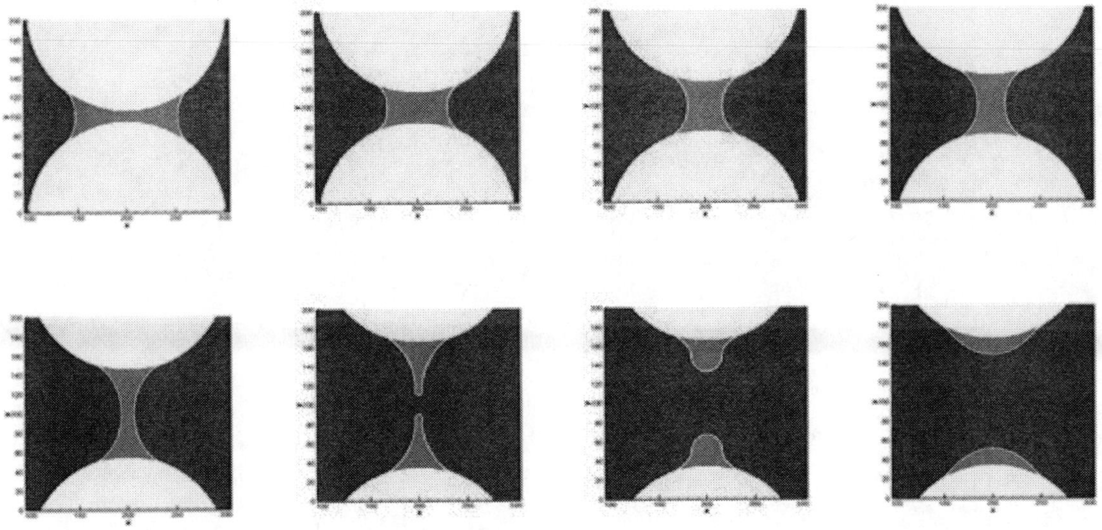

Figure 7 Snapshots of snapshots of dynamic liquid outlines at a given liquid volume

3.3 The tensile strength of a liquid bridge with shear deformation.

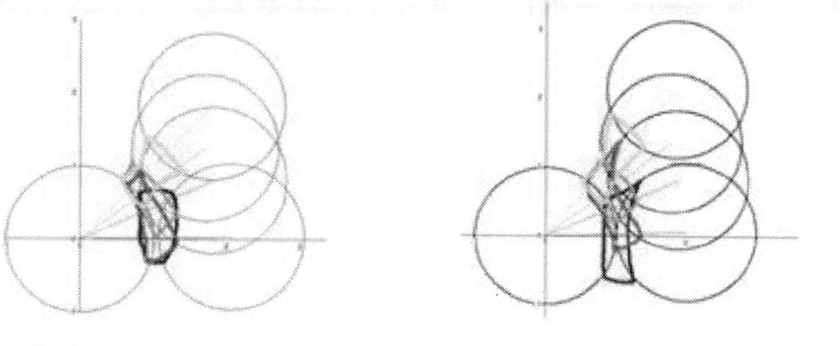

(a) Static state (G=0, 0.59, 0.98) (b) Dynamic state (G=0, 0.59, 0.98)

Figure 8 Liquid outlines at static and dynamic state (Bo=1.21, V=0.088)

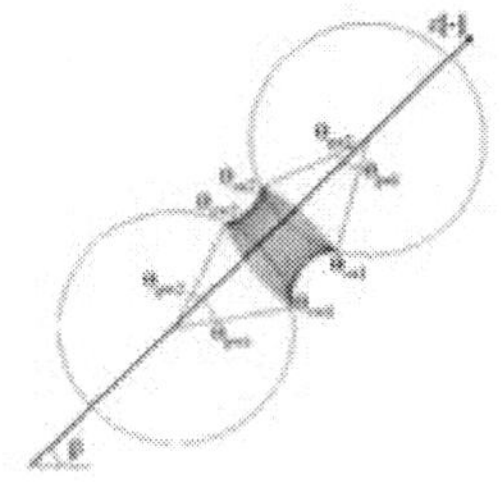

Definitions of half-filling and contact angles.

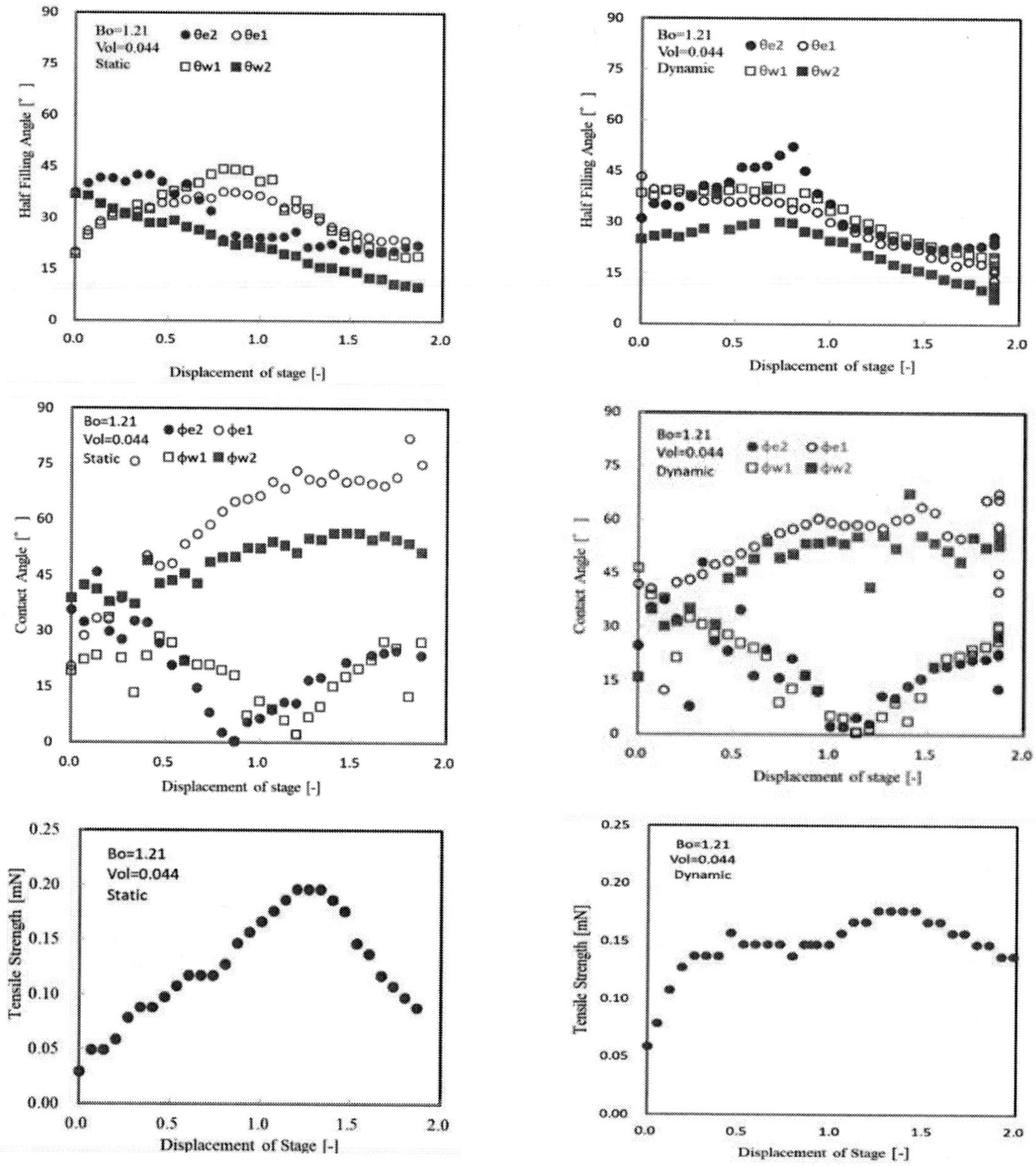

(a) Static state (b) Dynamic state

Figure 9 Profiles of half-filling angles, contact angles and tensile strength , Bo=1.21, V=0.088

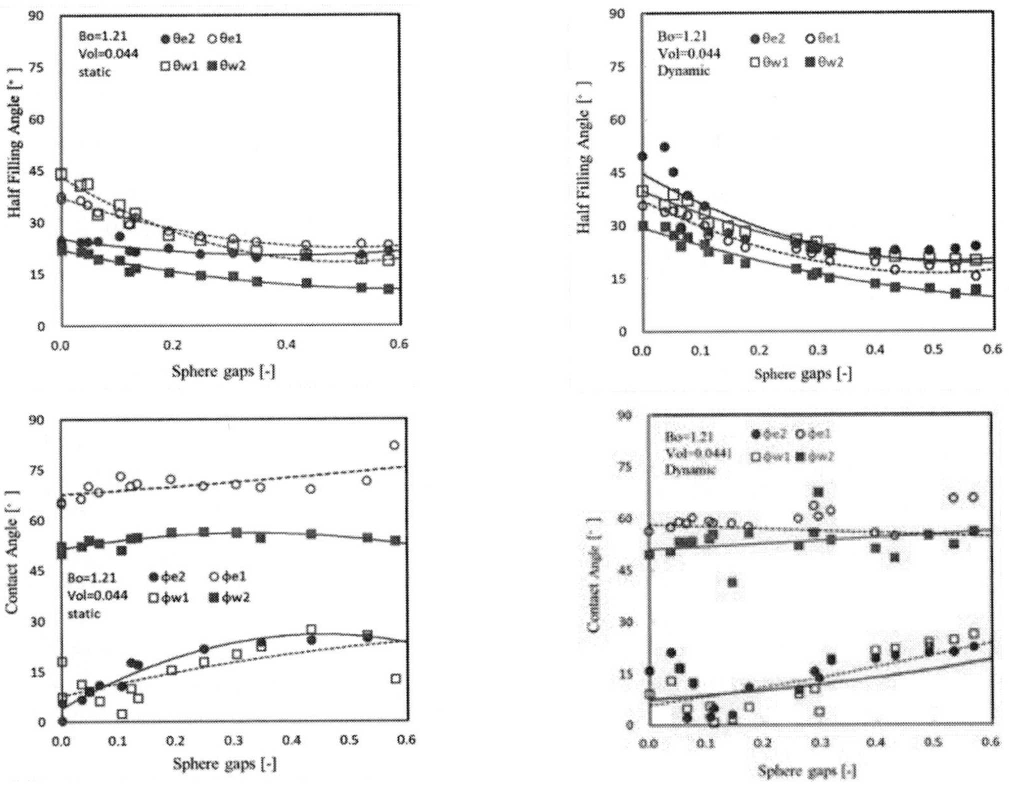

(a) Static state (b) Dynamic state

Figure 10 Profiles of half-filling angles, contact angles with sphere gaps , Bo=1.21,V=0.088

Figures 8 show four types of liquid outlines at static and dynamic state. Magenta and green plots are drawn respectively at the extreme values of half filling angles, contact angles and tensile strengths in Figures 9. During the lower stage displacement two spheres keep at the contact state and the half-filling angles reached to the peak values. Half-filling angles sifted with displacements of a stage have the peak values except the static half-filling angle at the top point of the lower sphere (θ_{Pw2}), which monotonously decrease with stage displacement. The increase and decrease of contact angle respectively corresponds to the advancing and receding contact angle. The contact angles at the bottom point of the lower sphere (θ_{Cw1}) and at the top point of the upper sphere (θ_{Ce2}) transit from the receding to the advancing contact angle due to the infinitesimal values. Figures 10 redraw the half-filling and contact angles with respect to sphere gap. Contact angles contribute to the peak tensile strength, which is measured as resultant force (F_T) of normal (F_n) and tangential (F_t) tensile strength in Figure 11. The normal tensile strength conventionally corresponds to the capillary force. The stage displacement at the peak tensile strength minimize the sum of half-filling and contact angle due to the positive hydrostatic pressure across liquid surface. The larger dynamic tangential strength occurs during the sphere contacting due to the asymmetric deformation. The adhesion force of liquid to sphere at the static state contribute to the larger peak tensile strength.

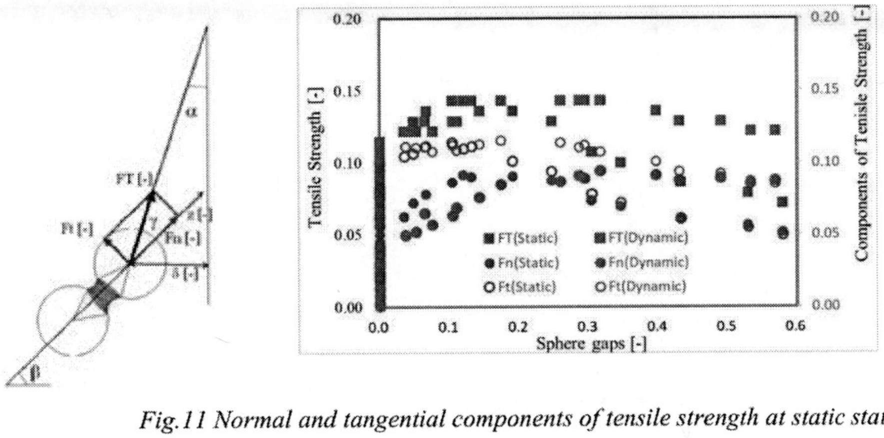

Fig.11 Normal and tangential components of tensile strength at static state

3.4 Numerical simulation at the static state

The solution of Young-Laplace equation demands for one unknown necessary condition on the contact point, i.e. half-filling or contact angle. Liquid surfaces simulated by FEM are drawn with solid lines in Figures 12. The simulation reveals the asymmetric liquid geometry with half-filling and contact angles along longitudinal angle. Disagreements of liquid interfaces results from the approximation of contact lines on spheres. The tangential strength influences depend on the mean curvature. Especially the second figure at the peak half-filling angle has the negative hydrostaic pressure in the direction of relative sphere displacement and positive pressure in the oppsite direction.

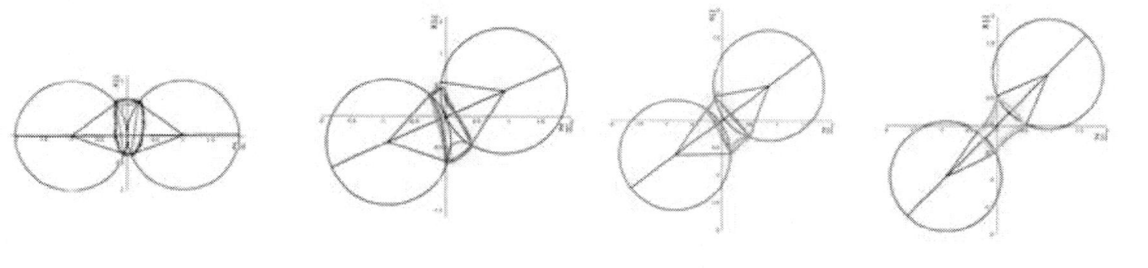

At the initial geometry At the peak halffilling angle At the peak tensile strength At the maximum shpere gap

Figure 12 Numerical estimation at the static state

Figure 13 displays transitions of measured normal tensile strengths ($F_{exp,n}$) and simulated capillary force (F_{cal}) to sphere gap. The initial droplet is attached to the above aperture in the direction of sphere movement due to gravity. The stage displacement drifts parts of liquid volume into the lower aperture of spheres. The sphere on displacement stage keep contacting with the other sphere hang up to the tensiometer up to the peak tensile strength. Fcal and FcalB are numerically estimated by outlines of liquid surfaces respectively with Young-Laplace equation and boundary toroidal model. The sign of hydrostatic pressure with deformation of liquid droplets results in the repulsive capillary force. The normal tensile strength corresponds to the capillary force during the range of symmetric liquid bridge. The angle of beta is defined by the inclination of center distance of a sphere pair. The lower normal tensile strength than the capillary force results from the orthogonal component of capillary force at the low range of beta.

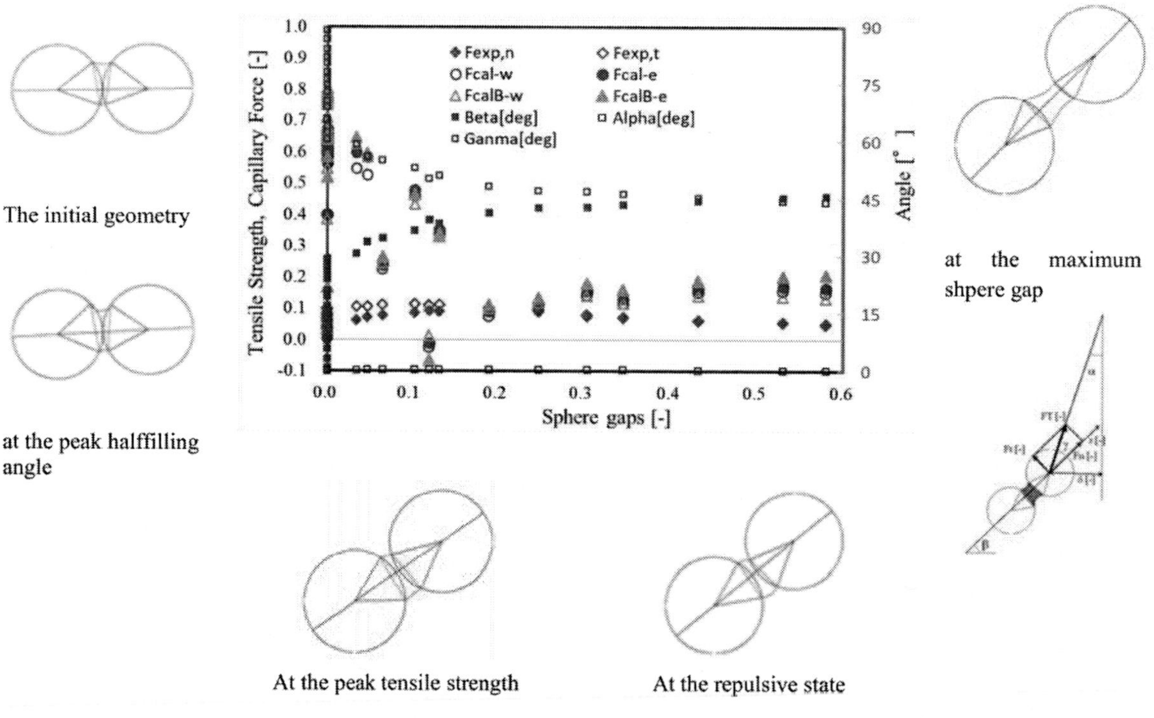

Figure 13 Normal tensile strength and numerical capillary force

Conclusions

Liquid bridges adhered to particles are deformed with the normal and tangential stress in wet granulator. The interparticle force with asymmetric droplet by shear stress must be estimated by the three dimensional solution of Young-Laplace equation. A surface tensiometer and numerical simulation contribute to the normal and tangential tensile strength of liquid bridge adhered to two spheres.

1. The normal strength of symmetric liquid bridge with a surface tensiometer is verified by the conventional toroidal model and the solution of Young-Laplace equation.
2. Sphere gaps at the peak capillary force in the vertical sphere configuration depend on liquid volume and displacement velocity of sphere.
3. The lattice Boltzmann Method is expected to numerically predict the boundary conditions on the contact lines, which are half-filling angle and contact angle at the static and dynamic state.
4. The tensile strength depends on transition of contact angles, which correspond to advancing or receding angles. The larger sum of contact and half-filling angles than the orthogonal angle corresponds to the positive hydrostatic pressure across liquid surface.
5. The normal tensile strength is related with the solution of Young-Laplace equation by FEM. The tangential tensile strength depends on the liquid transition within droplets and the variable mean curvature.

References

[1] K.Murase, H.Shibata, Y.Yasuda and S.Toyama, *Jpn J. Multiphase Flow,* **5**, 322(1991)
[2] A.E.Saez and R.G.Carbonell, *J. Colloid Interface Sci.,* **140**, 408(1990)
[3] D.N.Mazzone, G.I.Tardos and R.Peffer, *Powder Tech.*, **51**, 71(1987)
[4] C.D.Willet, M.J.Adama, S.A.Johnson and J.P.K.Seville, *Langmuir*, **16**, 9405(2000)
[5] M.J.Adama, S.A.Johnson, J.P.K.Seville and C.D.Willet, *Langmuir*, **18**, 6180(2002)
[6] R.K.Niven, *J Petroleum Sci. Eng.*, **52**, 1(2006)
[7] O.Pitois, P.Moucheront and X.Chateau, *Eur. Phys. J.B*, **23**, 79(2001)
[8] S.J.R.Simons, J.P.K. Seville and M.J. Adams, *Chem. Eng. Sci.*, **49**, 2331(1994)
[9] C.D.Willet, M.J.Adama,S.A.Johnson and J.P.K.Seville, , *Powder Tech.*, **130**, 63(2003)
[10] S.M.Iveson, J.A.Beathe and N.W.Page, *Powder Tech.*, **127**, 149(2002)
[11] R.A.Fisher, *J. Agri. Sci.*, 16,492(1926)
[12] F.M.Orr, L.E.Scriven and A.P.Rivas, *J. Fluid Mech.*, **67**, 723(1975)
[13] O.Pitois, P.Moucheront and X.Chateau, *J. Colloid Interface Sci.,* **231**, 26(2000)
[14] U.El Shamy and T.Groeger, *Int. J. Numer. Anal. Meth. Geomeh.*, **32**, 1763(2008)
[15] F.M.Orr, R.A.Brown and L.E.Scriven, *J. Colloid Interface Sci.,* **60**, 137(1977)
[16] K.Murase, T.Mochida, Y.Sagawa and H.Sugama, *Adv. PowderTech.,* **19**, 349(2008)
[17] C.D.Willet, M.J.Adama,S.A.Johnson and J.P.K.Seville, *Handbook of powder technology*, **11**, chap.28, ed. A.D.Salman, M.J.Hounslow and J.P.K.Seville,Elsevier (2007)

NUMERICAL SIMULATION OF WIRE-PLATE ELECTROSTATIC PRECIPITATOR - EFFECT OF PARTICLE CONCENTRATION

Ding Yang[1], Baoyu Guo[2], Jun Guo[1], Aibing Yu[2]*

[1] *Fujian Longking Co. Ltd, Longyan, Fujian, 364000, China*
[2] *Department of Chemical Engineering, Monash University, Clayton VIC 3800, Australia*

Introduction

Electrostatic Precipitators (ESP) have been widely used to separate fine particles from large scale industrial exhaust gases, such as coal fired power plants. In-depth knowledge and careful design optimization are critical to meeting the increasingly strict emission standard in terms of $PM_{2.5}$ emission limit, e.g., 5-10 mg/Nm^3. Numerical modelling has been employed as a powerful tool in this area. Complicated interactions among electric field, gas flow and particle transport are summarized by Adamiak [1]. To model such a system, various levels of simplifications have been made in published papers.

Among them, Lagrangian particle tracking technique is a widely accepted approach. The mass concentration of ash particles in coal-fire flue gas is typically 10-50 g/Nm^3, corresponding to volume fraction in the order of 10^{-5}. Such a small value is often regarded as sufficiently dilute particle concentration in numerical modeling of multiphase flows. For the case of ESP, complicated interactions among the e-field, gas flow and charged particles are involved, even if the deposited dust layer is ignored. Full coupling between the continuous gas phase and discrete particle phase is computationally expensive to achieve a converged steady solution. To avoid such a high cost, the model for ESP was simplified as so called one-way coupling in many previous works, so that the electric field is solved without the particle phase (Guo et al. [2,3]; Farnoosh et al. [4]). This simplification may be acceptable in the end stage of ESP, but will introduce significant error in the early collection stage. Limited studies reported in the literature have considered the particle charges as a source term in the Poisson's equation of electrostatics. For example, Choi and Fletcher [5] predicted a reduced current for the particle-loaded case (d_p = 1-5μm) compared with the particle-free case, but the corona ion charge density somehow shows an increasing trend along the stream. When fine dust particles are treated, the large specific surface area of the particle phase, combined with reduced collection efficiency, implies high space charge density in the collection region. Its effect was understood as a partial suppression of the corona current [6]. As was experimentally proved by Podlinski et al. [7], space charge of charged particles can contribute to the EHD flow pattern and particle transport. In present work, particle space charge is considered through two-way coupling and interesting results are generated in terms of ESP performance as particle loading changes.

Integrated ESP Modelling

Mathematical Equation

Coupled electric field, gas-particle flow and particle charging processes are simulated using

Eulerian-Lagrangian method in Computational Fluid Dynamics (CFD). Major governing equations of vector format are listed below for completeness. For electric field, Poisson equation of potential

$$\nabla^2 V = -(\rho_{ion} + \rho_c)/\varepsilon_0 \quad \mathbf{E} = -\nabla V \tag{1}$$

Current continuity equation

$$\nabla \bullet ((b\mathbf{E} + \mathbf{U}_f)\,\rho_{ion} - D\nabla\rho_{ion}) = S_c \tag{2}$$

where ρ_{ion} is corona charge density, b the ion mobility, D is ion diffusivity, ρ_c is particle space charge density, evaluated by averaging the charges of particles that pass through a control volume (cell),

$$\rho_c = \frac{1}{V_{cell}} \sum_{n=1}^{N} \dot{n}_p q_p \Delta t \tag{3}$$

Where $\dot{n}_p$ is the number flow rate of a parcel of particles tracked, Δt is the residence time in the cell. The source term, S_c, in Equation (2) is transfer rate of ion charges to particles, which is not considered at the moment, i.e., $S_c = 0$. For the gas flow, the momentum equation is,

$$\begin{aligned}\nabla \bullet (\rho_g \mathbf{U}_g \otimes \mathbf{U}_g) = \nabla(p + \tfrac{2}{3}\rho_g k) + \nabla \bullet [(\mu + \mu_t)(\nabla \mathbf{U}_g + (\nabla \mathbf{U}_g)^{\mathrm{T}})] + \rho_{ion}\mathbf{E} \\ - \frac{3\pi n_p d_p \mu}{\mathrm{C_C}}(1 + \mathrm{Re}_p^{0.687})(\mathbf{U}_g - \mathbf{U}_p)\end{aligned} \tag{4}$$

For the particle flow, the force balance of a single particle is

$$m_p \frac{d\mathbf{U}_\mathrm{p}}{dt} = \frac{3\pi n_p d_p \mu}{\mathrm{C_C}}(1 + \mathrm{Re}_p^{0.687})(\mathbf{U}_g - \mathbf{U}_p) + \tfrac{1}{6}\pi d_p^3 \mathbf{g}(\rho_p - \rho_g) + n_e q_e \mathbf{E} + \mathbf{F}_{dis} \tag{5}$$

k-ω SST turbulence model is chosen for gas flow, and the dimensionless Lawless charging rate equation is used for particle charging [8].

Model Conditions

An integrated ESP model developed comprises two parallel modules or submodels, namely, electric field model and coupled gas-particle flow model. One-way coupling between the two modules is used earlier is upgraded to two-way in the present work. The electrostatic field caused by charged particles in suspension is considered in the solution of electric field. The two submodels are coupled through a small under-relaxation factor, in views of strong interaction between the electric field and particles flow, and the stochastic nature of the discrete particle phase. In this way all the volume averaged quantities related to the particle phase (such as the particle space charge density) are smoothed out over time in the simulation.

A single channel, 4-wire and flat plate configuration is considered. The dimensions are 3.5mm in wire diameter, 0.24m for wire-wire spacing, 0.2m for wire-plate spacing. The plate length is 1.2m. The simulation domain is defined as 1.74m long by 0.4m wide box with the 4 holes inside. The grid is smoothly refined at places where higher gradient in electric quantities is expected, such as the proximity of the wires and middle of the adjacent wires. Other base case conditions are as follows.

- Applied voltage on the wire electrode: Va = -60kV. The charge density on the wire surface is determined using an injection law combined with Peek equation (Guo et al.[9]).
- Gas temperature: 120°C. As the inlet boundary condition for the gas phase, the velocity is set as 1 Nm/s; relative turbulence intensity 3.7%, turbulent length scale $0.3D_h = 0.24$m.
- Dust particle size is mono-dispersed at d_p=2.5 μm. Particle density 2000 kg/m^3; ash dielectric constant of κ_p=6.6, Particle loading at the inlet, $(\rho_p)_{in}$, is varied with default value of 10 g/Nm^3.

Typical Results

Electric Field without Particles

When the inlet/outlet boundaries are moved sufficient long distance away from the electrodes in the simulation domain (referred to as "natural boundary condition"), the electric field around the boundary wires show an asymmetric distribution pattern that differ from that around the inner wires. Electric field-lines/current can go out of the limit of the collection region (Figure 1). Electric field intensity on the collection plate is of interest as it is considered a factor to determine the upper limit of operation before spark-over [10]. Normally the field intensity on the plate is highest at the point that faces exactly the wire center with the least distance. This maximum field intensity on the plate is 3.80×10^5 V/m. However, when the natural boundary condition is used, the highest value (1.06×10^6 V/m) predicted in field intensity appears at the endpoint of the plate due to its sharpness. This is a likely point that would trigger electric spark-over prematurely. Practically such a sharp edge should be avoided. Generally, the field strength at the middle wires is smaller than that at a boundary wire, due to the counteraction of adjacent wires. Then the corona charge density at the boundary wires should be greater than that at the inner wires (Figure 2), which is similar to the simulation result of Kasdi [11].

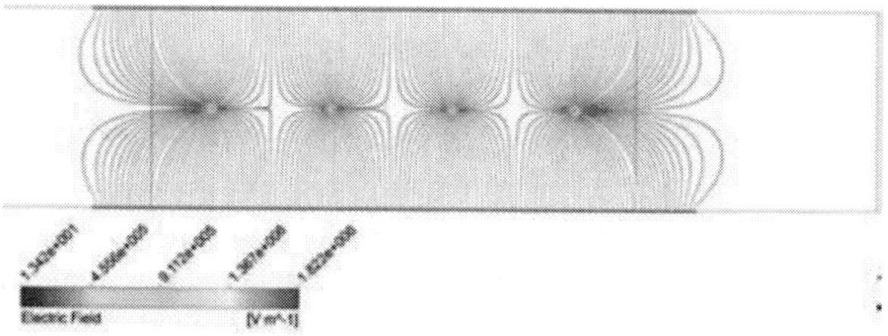

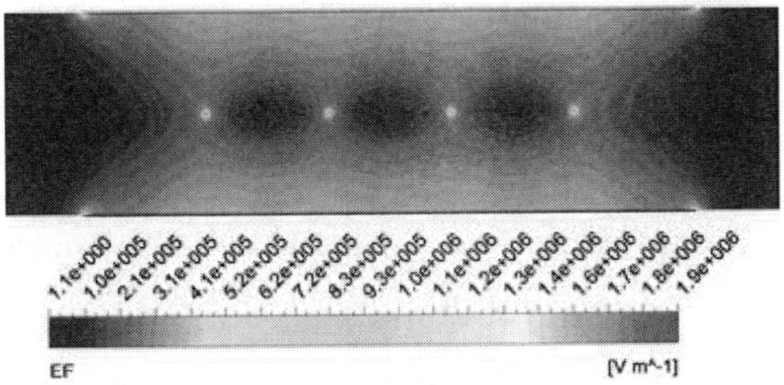

Figure 1. Field streamlines and intensity distribution (no particles).

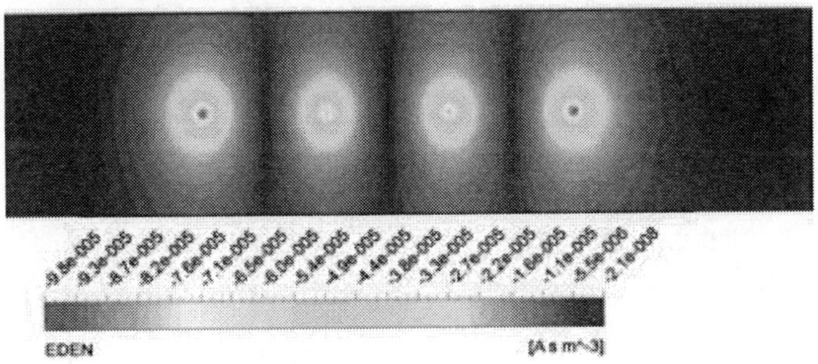

Figure 2. Corona charge density (no particles).

Effect of Particle Space Charge

Particles are randomly injected from the inlet plane with uniform probability in locations and no-slip for velocity (same as the gas velocity). In using random number generator, it is ensured that the injection points are not repeated for every coupling cycle. Figure 3 shows sampled particle trajectories with color scaled to particle charge. Figure 4 shows a pattern of particle deposition rate on the plate. It is seen that the end points of the plate, particularly the downstream one, are special spots to accumulate more dust.

Figure 3. Particle trajectories with colour scaled to particle charge.

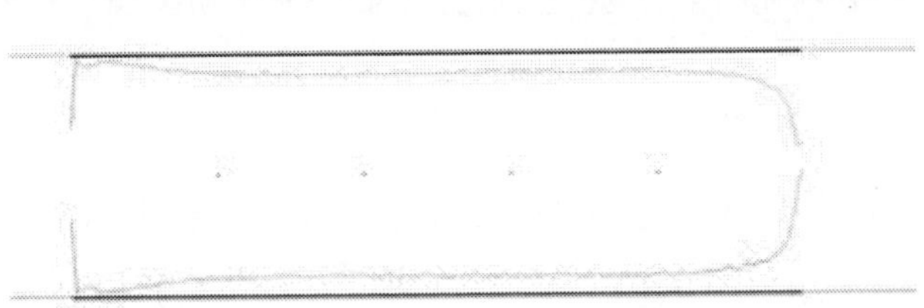

Figure 4. Indicative pattern of particle deposition rate on the plate.

Figure 5 shows the averaged space density of charges carried by particles. Highest space charge density appears near the first wire electrode, where electric current density is highest. It diffuses gradually from the centre line with particles drifting toward the plate while moving forward. Due to the radiating nature of the field-lines around each wire, the lateral component of Coulomb's force in the region close to the centre line may not be as large as that off the centre line. Not collected charged particles of negative polarity are conveyed towards the outlet, lowering the potential downstream outside the collection region (Figure 6), causing distortion of the overall electric field distribution.

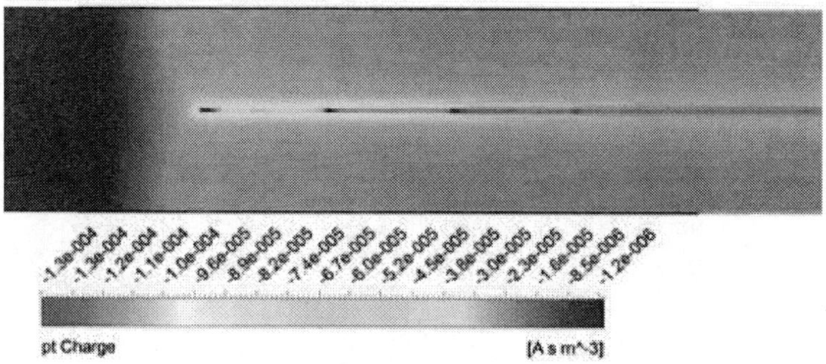

Figure 5. Distribution of space charge density of particles.

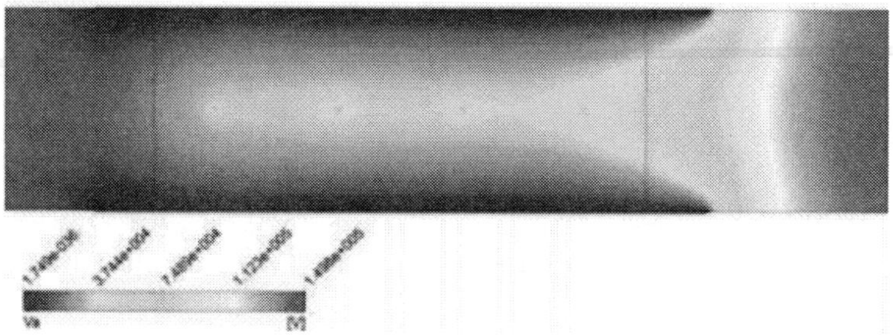

Figure 6. Distribution of electric potential (in absolute value).

Effect of Particle Loading

On Electric Current

Figure 7(a) shows the mean corona current, based on unit length of wire. As can be seen, the corona current drops quickly as particle loading increases, although a fixed voltage is applied. This is caused by the secondary electrostatic field created by the space charges of particles that cancels out the primary electric field at the wire surface and subsequently suppressed the corona generation.

On Particle Collection Efficiency

Predicted particle collection efficiency, as particle loading changes, is given in Figure 7(b). The collection efficiency, increases initially with particle loading (<3 g/m^3), peaks at particle loading of 4-5 g/m^3 and then decreases as particle loading increases further. When ignoring particle space charge, the electric field would be independent of particle loading. This is the case for very low particle loading, where the fractional collection efficiency is constant, or in other words, the quantity of dust collected as well as that penetrated increases proportionally with particle loading.

The observed enhancement of collection by a slight amount of particles is due to the redistribution of electric field so that the near-plate and volume-averaged field intensity happens to increase, which favor the charging of most particles. The asymptotic lower limit of the curve should correspond to the situation when the corona current is so small that turbulent dispersion dominates over the drift force.

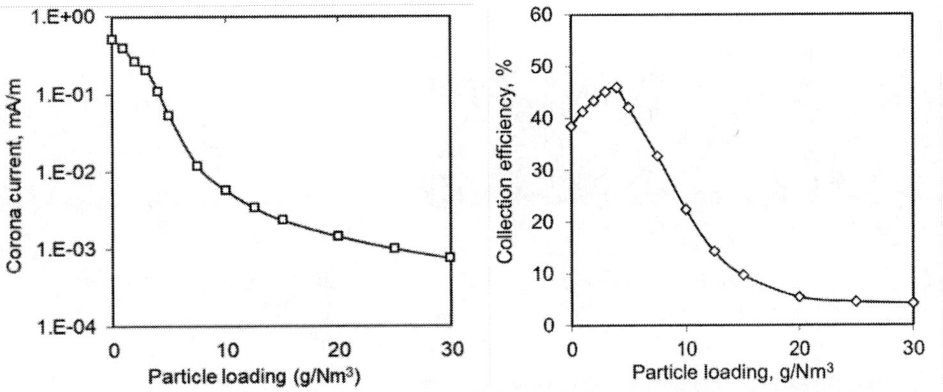

(a) (b)

Figure 7. Mean corona current (a) and collection efficiency (b) as function of particle loading.

On Electric Field Distribution

Figure 8 visualizes the predicted corona charge density and field-lines, which shows how the electric field distribution evolves with particle loading. As the electric field is distorted starting first from the end side, corona density at 3 of the 4 downstream wires is suppressed faster than the 1st wire. When the particle loading is above 7.5 g/m^3, all wires except the first one stop emitting. Meanwhile, reversed field-lines appear around the last wire, and the affected region expands as the loading increases further. Otherwise the field-lines become smoother with increasing loading. Since the field-lines are extended to the outlet, some escaped particles can be pulled back into the collection region, a factor that contributes to significantly greater deposition rate around the end-point of the plate.

Overall, the particle charges affect the particle collection adversely, as they tend to suppress the corona emission or reduce the corona current. Particularly the penetrated particles, which carried significant charges, have prolonged effect as they float along. The corona at the downstream electrodes even extinguish unless the applied voltage is raised individually. When highly charged, more particles precipitate to the plate, reducing the number of particles floating ahead (or less space charge density). The charges carried by these remaining particles exert a negative feedback to the electric field and particle charging, which compromises the collection efficiency.

Electric Quenching

Corona quenching is a common phenomenon in ESP when fine particles are treated. Riebel et al. [12] experimentally studied the corona discharge characteristics of ESP with glycerol aerosol, which found corona onset voltage increases and the corona current significantly decreases after the glycerol steam was injected into the system. Under fixed voltage in the present simulation, corona currents from all the wire electrodes drop with increasing solid loading. The downstream coronas (from the second one) will eventually extinguish at some point. An interesting phenomenon arises when the particle loading increases above 10 g/Nm^3, i.e., the electric streamlines in a small zone around the last wire reverse direction locally. This implies that a charged particle, once falls accidently within the zone, could be trapped and attracted to the same wire, as seen from the particle trajectories in Figure 9. Thus this wire is subject to heavier dust deposition.

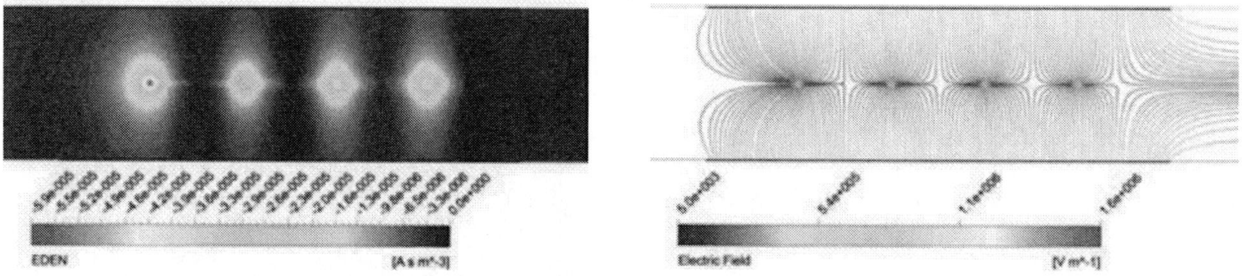

$(\rho_p)_{in}$ = 3 g/Nm^3

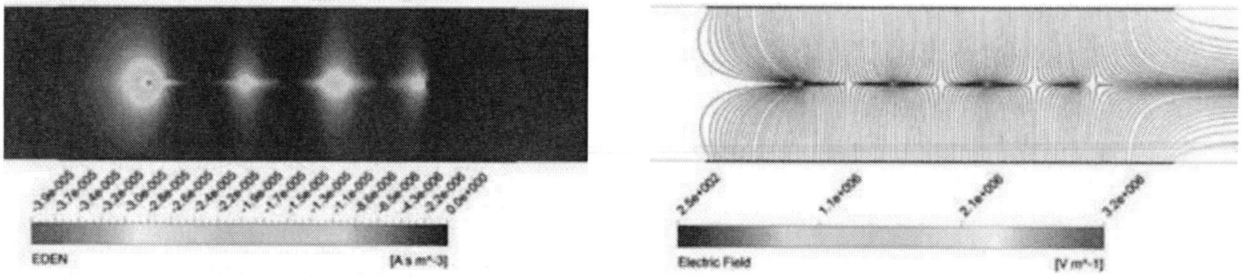

$(\rho_p)_{in}$ = 5 g/Nm^3

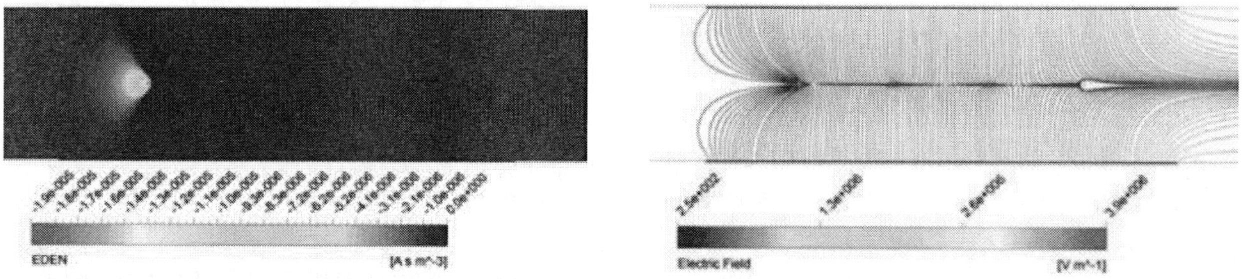

$(\rho_p)_{in}$ = 10 g/Nm^3

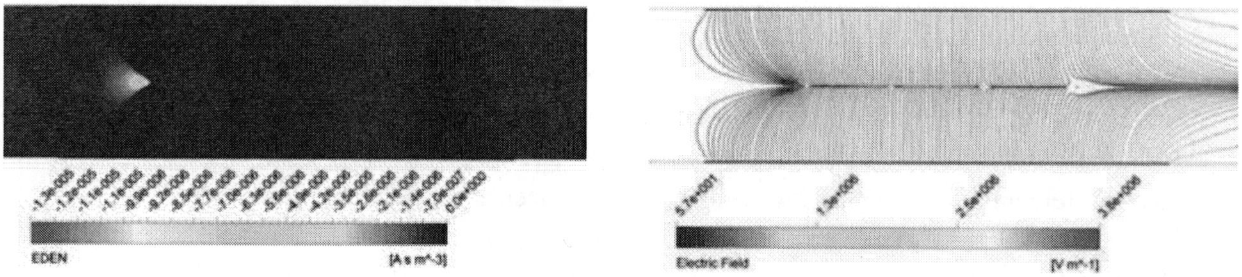

$(\rho_p)_{in}$ = 20 g/Nm^3

Figure 8. Visualizations of corona charge density and field streamlines for different particle loadings.

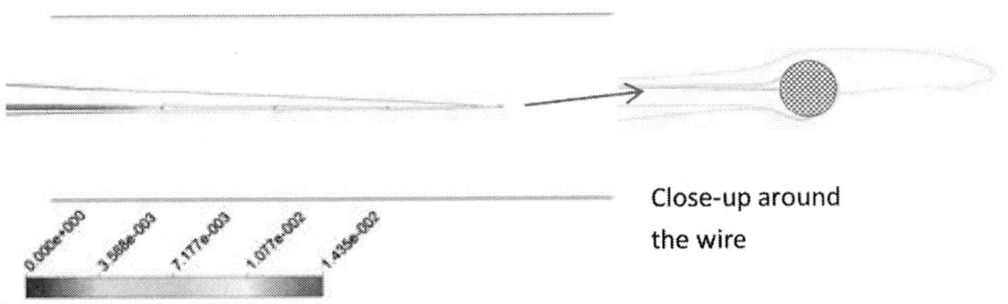

Figure 9. Trajectories of particles falling on the corona wire ($(\rho_p)_{in}$ = 20 g/Nm^3).

Concluding Remarks

CFD based ESP model has been advanced, with an enhanced coupling between electric field and particle flow. So the model can be readily applied to the early collection stage of the ESP, where dust particle loading is too high to neglect. Space charge of particles is considered, and its significance is clarified in a broad range of particle mass concentration in the gas. Dust loading significantly affects the dust collection efficiency for fine particles considered. Maximum collection efficiency exists at a critical particle loading that depends on particle size. However, as particle loading increases, collection efficiency decreases and eventually approaches asymptotically to a lower limit. The phenomenon of electrical quenching is observed, where downstream corona is extinguished or one of the emitters is polluted when the particle loading is high.

References

1. Adamiak K. (2013), "Numerical models in simulating wire-plate electrostatic precipitators: A review", *J. Electrostatics*, 71, pp.673-680.
2. Guo B.Y., Yu A.B. and Guo J. (2014), "Numerical modelling of electrostatic precipitation: effect of gas temperature", *J. Aerosol Science*, 77, pp.102-115.
3. Guo B.Y., Yu A.B. and Guo J. (2015), "Numerical modelling of ESP for design optimization", *Procedia Engineering*, 102, pp.1366-72.
4. N. Farnoosh, K. Adamiak, G.S.P. Castle (2010), "3-D numerical analysis of EHD turbulent flow and mono-disperse charged particle transport and collection in a wire-plate ESP", *J. Electrostatics*, 68, pp.513-522.
5. Choi B.S. and Fletcher C.A.J. (1998), "Turbulent particle dispersion in an electrostatic precipitator," *Applied Mathematical Modelling*, 22, pp.1009-1021.
6. Parker K.R., Applied Electrostatic Precipitation, Blackie Academic & Professional, 1997.
7. Podlinski J., Niewulis A., Mizeraczyk J. and Atten P. (2008), "ESP performance for various dust densities", *J. Electrostatics*, 66, pp.246-253.
8. Lawless P.A. (1996), "Particle charging bounds, symmetry relations, and an analytic charging rate model for the continuum regime", *J. Aerosol Sci.*, 27, pp.191-215.
9. Guo B.Y., Guo J. and Yu A.B. (2014), "Simulation of the electric field in wire-plate type electrostatic precipitators", J. Electrostatics, 72, pp.301-310
10. White H.J. (1974), "Resistivity problems in electrostatic precipitation", *J. Air Pollut. Control Assoc.* 24, pp.313–338.
11. Kasdi A. (2016), "Computation and measurement of corona current density and V–I characteristics in wires-to-plates electrostatic precipitator", *J. Electrostatics*, 81, 1-8.
12. Riebel U., Radtke R. and Loos R. (2002), "An experimental investigation on corona quenching", *J. Electrostatics*, 54, pp.159-165.

Triboelectric Charge of Spherical Glass Particles against Metal Pipeline

HoSu CHOI [1], Teruo SUZUKI [2] and Kwangseok CHOI [3,]*
[1] Christian Academy in Japan, [2] Kasuga Denki Inc.,
[3] National Institute of Occupational Safety and Health, Japan
**choiks@jniosh.go.jp*

Introduction

Insulating powders are used in many ways such as in the food, pharmacy, and chemical industries. Usually, these powders are highly charged when transported, loaded, dried, and mixed, due to repeated contact and separation between powder particles and the equipment. This results in the powders acquiring electrostatic charges. If not dissipated quickly, electrostatic charge buildup can cause inter-particle cohesion and adhesion to operating equipment, causing a production efficiency decrease. Incendiary electrostatic discharges such as spark discharges, cone discharges, and propagation brush discharges can also take place during industrial operations and may lead to fires and even explosions. Riedewald reported on an industrial fire in a pharmaceutical powder transfer operation [1]. A large number of studies involving electrostatic charge generation on powders, as well as electrostatic discharges and hazards during operation, have been reported. For example, Kaialy reviewed the main factors affecting the electrostatic charging of pharmaceuticals such as particle resistivity, size, shape, and surface roughness [2]. Gibson also reviewed industrial hazards related to electrostatic charges and discharges [3]. Our previous studies [4–6] investigated the electrostatic charging of polymer powder in mixing processes and the ignitability of aluminous, magnesium and polymer powders due to electrostatic discharges involving the effect of the ratio of N2 in the air on the ignitability.

In order to easily measure and confirm the electrostatic charge level of powders, a novel tribocharging system that utilizes the Faraday cup is presented, and the charging behavior of the sample powders is investigated in this paper.

Experimental Apparatus and method

Figure 1 shows the spiral air tribocharge apparatus presented in this study. The apparatus was divided into four main parts: the air supply part, the powder supply part, the tribocharge part, and powder collection part. The air supply part consisted of a compressor (Hitachi, PO3.7MA) that included an auto dryer (CKD, F3-217199), air controller, electronic valve, and switch. In this study, dried air was used because moisture in the air may have an effect on the electrostatic characteristics of the powder sample. The powder supply and tribocharge parts consisted of a powder port, a spiral air flow device, and a pipeline. The spiral air flow device (photo 1) was attached in front of the sample powder port in order to create spiral air inside the pipeline. The spiral air flow device had 8 holes that were 6 mm in diameter and that opened at a slant (60 degrees), which is very different from conventional models [7, 8, 9]. The straight pipeline was made up of stainless steel metal and was 1300 mm in length and 35 mm in diameter. In order to prevent external noise, the overall powder supply and tribocharge parts were shielded. The powder collection part was made up of a cyclone-type Faraday cup, unlike conventional models [7, 8, 9]. The cyclone-type Faraday cup had a separate metal container (30 mm in diameter) for easy measurement of the amount of collected powders. The other devices included two electrometers (Kethley,

6517A), a data logger (Graphtec, GL900), an electrostatic ionizer (Kasuga Denki Inc., KD-510W), an electrical balance (Shimazu, BL220H), and a timer.

The typical operating procedure for the tribocharge test is summarized as follows:
(1) Adjust the air pressure using the air pressure valve.
(2) Load 1 g of the sample powder into the powder port.
(3) Eliminate the initial charge of the sample powders using an electrostatic ionizer.
(4) Reset the electrometers (Q mode) connected to the pipeline and cyclone.
(5) Switch the air flow on, and run the sample powder through the pipeline where friction occurs.
(6) Measure the indication value when it stabilizes.
(7) Measure the amount of powder collected in the Faraday cup.
(8) Clean out the pipeline and cyclone part using an air spray in preparation for the next test.

The charge to mass ratio, q (nC/g), of the sample powder was obtained by dividing the charge, Q (nC), by mass, m (g), of the sample powder loaded into the Faraday cup:

$$q = \frac{Q}{m} \quad \cdots\cdots (1)$$

This experiment is mainly to determine how much the charge to mass ratio, q (nC/g), of the sample powders is affected by the size of the powder and the amount of air pressure. The cyclone-type Faraday cup and the pipeline were suspended electrically with insulators. All experiments were conducted in ambient conditions (air temperature: 22 ℃; relative humidity: 55 %).

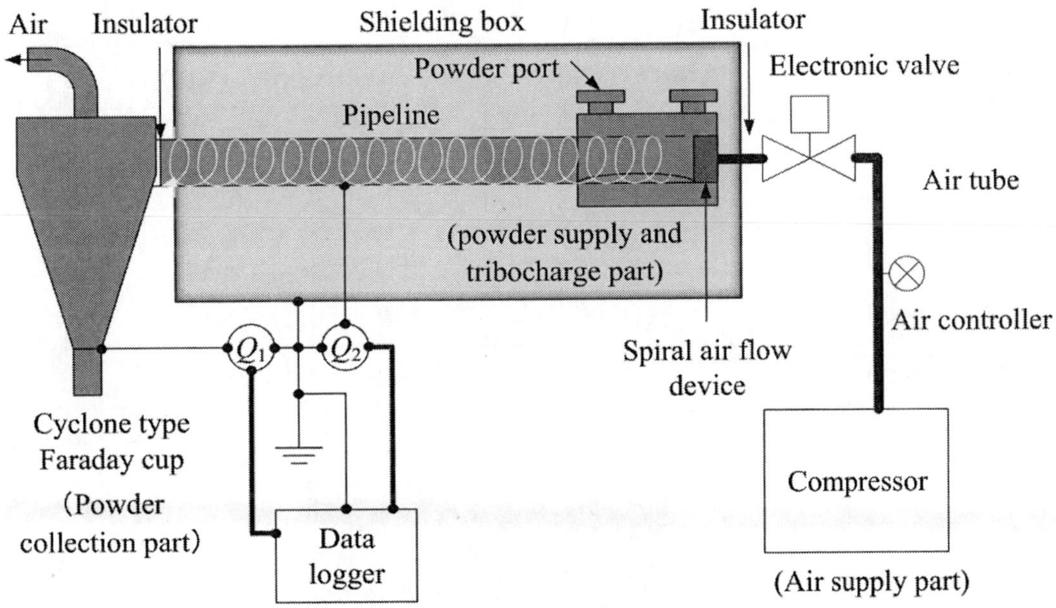

Fig. 1 Schematic of the spiral air tribocharge apparatus used in this study.

Photo 1 Structure of the powder port and spiral air flow device.

Sample powders

Glass beads are extremely good light reflectors and thus are widely used as the main component of the coating material for light-reflecting adhesive tape, background paint for traffic control sign boards, and paint for road signs. Four different grades of commercial glass beads were used in this experiment, as shown in Table 1. The particle size was determined using a laser diffraction system (LDA Win 1.21, dry type). Hereafter, they will be called by their abbreviation shown in Table 1. The analysis of the images from microscopy (Neoscope, JCM-500) revealed that the G-4 powder particles were spherical, as shown in photo 2. The other glass bead particles were the same shape as G-4.

Table 1. List of powder specimens used in experiments.

Specimen	**Abbreviation**	**Diameter [μm]**		
		D_{10}	D_{50}	D_{90}
Glass beads 1	G-1	22	62	85
Glass beads 2	G-2	93	131	175
Glass beads 3	G-3	130	180	241
Glass beads 4	G-4	162	225	294

D_X: X% volume average.

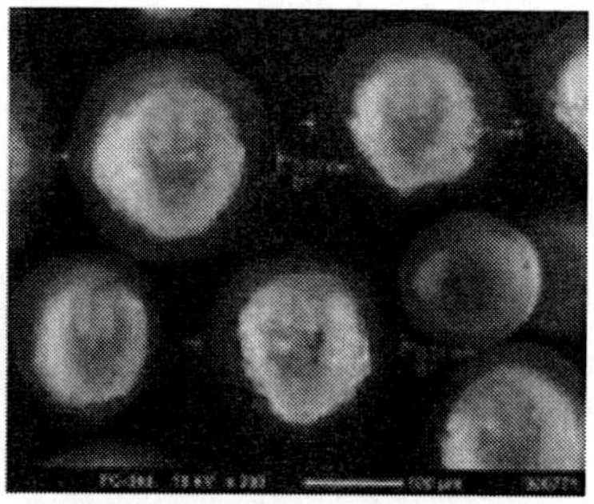

Photo 2 Shape of G-4 powder used in this experiment.

Experimental results and discussion

Figure 2 shows the typical change of charge detected over time when the G-3 particle was used and the air pressure was 0.15 MPa. It should be noted here that when no particle was used and only air was passed through the pipeline, the charge was detected as almost zero. The charge Q1 of the powder was found through the cyclone-type Faraday cup and then was confirmed through the charge Q2 obtained from the pipeline. As the powder particles passed through the pipeline for about 0.82 seconds, as seen in the Fig. 2, the charge clearly increased. Q1 and Q2 had almost the same trend and absolute values. This result confirmed the reliability of the test and showed that there was no disturbance by external noise.

In this study, all of the glass beads (G-1 to G-4) were negatively charged when stainless steel metal was used as the contact surface. The charge polarity of the materials relates to the work function (or ionization potential) of each material. The contact potential difference between two substances is defined by the difference between their fermi levels, which correlate with the work function. In general, materials with higher work functions tend to appropriate electrons from materials with lower work functions. This point will be carried out in our further study.

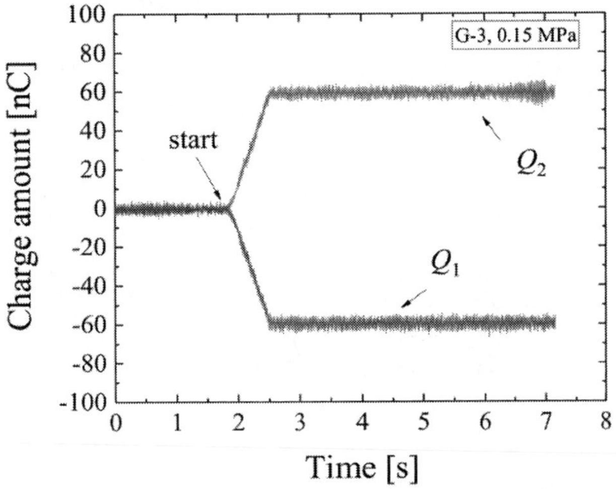

Fig. 2 Change of charge detected over time.

Figure 3 shows the charge to mass ratio of glass beads as a function of applied air. The range of applied air was between 0.1 MPa and 0.3 MPa in this study. When the applied air pressure was below 0.1 MPa, it was most likely not enough to send all of the powder through the pipeline. Each experiment was repeated five times, and the average values were reported. This graph does not contain an error bar, but these were about 10%. The results showed that the charge to mass ratio of both powders increased as the applied air increased. Although not shown in Figs. 2 and 3, the increase in air pressure also causes the slope of the graph to increase significantly. This can be explained by the fact that with increased pressure of applied air, the particles collide against the surface of the pipeline with more strength [9]. In order to investigate in more detail, the velocity of air at approximately 10 mm from the inside of the pipeline wall was measured using an anemometer (Kanomax, Anemomaster-6115). The approximate velocity of air flow was 9.5 m/s at 0.1 MPa, 12 m/s at 0.15 MPa, 15 m/s at 0.2 MPa, 20 m/s at 0.25 MPa, and 22 m/s at 0.3 MPa. It could be clearly seen that as the air pressure increased, the air velocity also increased. Another reason for this may be that because the air was applied in a spiral form, the particles may have collided more frequently against the pipeline when the air flow was increased. This exception will be investigated in detail in the future.

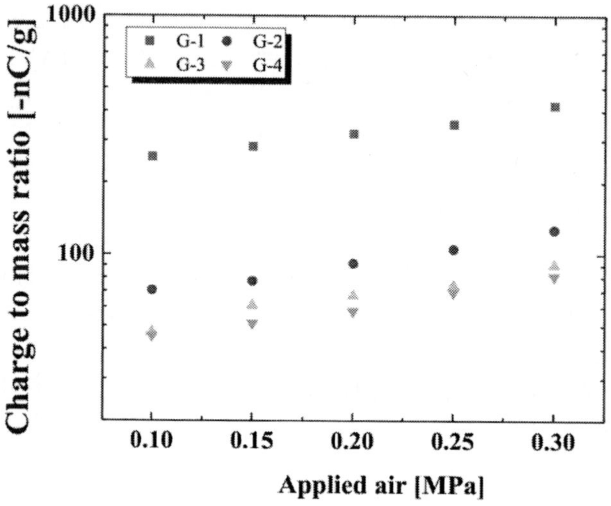

Fig. 3 Charge to mass ratio of glass beads as a function of applied air.

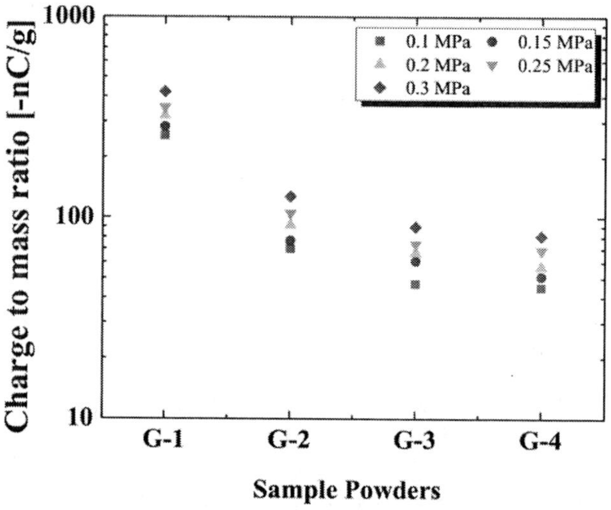

Fig. 4 Charge to mass ratio as a function of different glass bead sizes.

Figure 4 shows the charge to mass ratio as a function of different sizes of glass beads. The size distribution of each glass bead in this study was described in Table 1. The results showed that the charge to mass ratio of the glass beads increased as the powder size decreased. That is to say, smaller particles had a larger q than the larger ones. The total surface area of all of the smaller beads combined is greater than that of the larger ones, which results in a high charge to mass ratio development in the tribocharge. The value range of the q of the glass beads obtained in this study was about 40 nC/g to 100 nC/g. These values agreed with those obtained from the micronizing and pneumatic transfer process [10]. These values were high enough to generate incendiary electrostatic discharges during handling in industrial processes. This suggests that if glass beads are used with solvents in the cleaning process, they must be handled carefully, as incendiary discharges can ignite the solvents even though the glass beads themselves are nonflammable material.

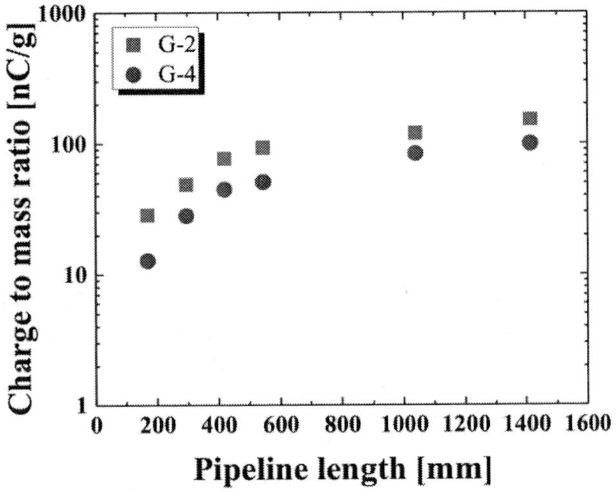

Fig. 5 Charge to mass ratio as a function of pipeline length.

Figure 5 shows the charge to mass ratio as a function of pipeline length. Six different pipeline lengths, *d*, were used: 170, 295, 420, 545, 1040, and 1415 mm. The glass beads used in this experiment were G-2 and G-4, and the air pressure was 0.35 MPa. Since the pipeline length changed with each test, it was difficult to attach the Faraday cup to the end of the pipeline. Therefore, in this study, only the value of q_2 was measured. As seen in Fig. 2, q_2 had the almost same value as q_1 but had an opposite sign. The length of the pipeline was measured starting from beginning of the powder port because the inside structure of the powder port was the same as that of the pipeline. Each experiment was repeated five times, and the average values were reported. The results showed that q_2 increased as the length of the pipeline increased for both sizes of glass beads. This can be explained by the fact that as the length increases, the time the glass beads come in contact with the pipeline also increases. The increase rate at the beginning section of the pipeline was especially large. It can also be seen through Fig. 5 that overall, G-2 had a larger q_2 than G-4 because of the total surface area. This confirms the reliability of the data shown in Fig. 4.

In addition, both beads seemed to reach saturation when the pipeline length was 1415 mm. It was compared to the maximum charge to mass ratio, q_c [nC/g] calculated from the well-known equation shown below [11]:

$$q_c = |\sigma_{max}|/(D \cdot \rho), \qquad \text{--------------------------(2)}$$

where σ_{max} [kg/m^3] is the maximum surface charge density (27 μC/m^2); D [μm] is the particle diameter; and ρ is the particle density (1.5×10^3 kg/m^3).

In this study, the particle sizes of G-2 and G-4 were 131 μm and 225 μm, respectively, as shown in Table 1. The calculated maximum values q_c and experimental maximum values q_t are compared as shown in Table 2. The calculated and experimental values for both G-2 and G-4 agreed well with each other. Therefore, it can be seen that both G-2 and G-4 reached saturation.

Table 2. Calculated maximum values and experimental maximum values of the charge to mass ratio.

	q_c [nC/g]	q_t [nC/g]
G-2	137	151
G-4	80	99

Conclusion

In this paper, a novel tribocharging system that utilizes the Faraday cup was presented and the electrostatic charge behavior of glass beads was experimentally investigated. The results were as follows:
1) For all of the glass beads, the charge clearly increased as the glass bead particles passed through the pipeline. The charge amount of the glass beads and the charge amount obtained from the pipeline had almost the same trend and absolute values. This result confirmed the reliability of the test and showed that there was no disturbance by external noise.
2) All of the glass beads were negatively charged when stainless steel metal was used as the contact surface. The value range of the charge to mass ratio of the glass beads obtained in this study was about 40 nC/g to 100 nC/g.
3) The charge amount of the glass beads increased as the applied air increased. It could also be seen that as the air pressure increased, the air velocity also increased.
4) The charge amount of the glass beads increased as the powder size decreased due to the total surface area of the beads.
5) The charge amount of the glass beads (G-2, G-4) increased as the length of the pipeline. The increase rate at the beginning section of the pipeline was especially large.

Acknowledgments
The authors gratefully acknowledge Mr. Naoto Nogera for his assistance during the test.

References

1. F. Riedewald (2012): "A fire in a secondary pharmaceutical powder transfer operation," *Process safety progress*, 31, 4, pp. 390-392.
2. W. Kaialy W, "A review of factors affecting electrostatic charging of pharmaceuticals and adhesive mixtures for inhalation," *Int.J.Pharm.* Vol.503, pp.262-276.
3. N. Gibson (1997), "Static electricity — an industrial hazard under control," *J. Electrostatics,* Vol. 40-41, pp. 21-30
4. K. Choi, K. Choi, K. Nishimura (2015), "Experimental study on the influence of the nitrogen concentration in the air on the minimum ignition energies of combustible powders due to electrostatic discharges," *J. of Loss Prevention in the Process Industries.* Vol.34.pp.163-166.
5. K.S. Choi, H.Sakasai, and K. Nishimura (2015), "Experimental study on ignitability of pure aluminum powders due to electrostatic discharges and Nitrogen's effect," *J. of Loss Prevention in the Process Industries.* Vol.35.pp.232-235.

6. K.S. Choi, H. Sakasai, and K. Nishimura (2015), "Minimum ignition energies of pure magnesium powders due to electrostatic discharges and Nitrogen's effect," *J. of Loss Prevention in the Process Industries.* Vol.41.pp.144-146.
7. N. Masui, and Y. Murata (1981), "charging tendency of the flowing powder through a metal tube (II), " *The five Annual meeting of Institute of Electrostatic Japan,* pp. 244-247.
8. S. Watano, S. Saito, and T. Suzuki (2002), "Feedback control of powder electrostatic charge in a pneumatic conveying process," *J. Soc. Powder,Japan*, Vol. 39, pp. 496-502.
9. P. Boschung, M. Glor (1980), "Methods for investigating the electrostatic behavior of powders, J. of elelctrostatic," Vol. 8, pp. 205-219.
10. IEC50404(2004), "Electrostatics- Code of practice for the avoidance of hazards due to static electricity," p. 66.
11. M.Glor (1988), "Electrostatic hazards in powder handling," *Research studies press LTD.*, p.26.

509266 Spray-Drying of a Layered Double Hydroxide Nanosuspension

Abstract:

Over the past few years, layered double hydroxide (LDH) has attracted particular interests in both academia and industry. LDH is 2D anionic clay of brucite-like layered structure with various organic or inorganic anions located between layers. The diverse composition of LDH, and its associated physical and chemical properties open up new possibilities in various applications. LDHs are successfully applied as precursors in synthesis of supported catalysts, sorbents, carriers delivering bioactive molecules, matrices containing active pharmaceutical ingredients for control release, electrochemical biosensors, etc. LDHs are mainly synthesized in a form of nanosuspensions. The application of nano-sized sorbents or catalysts in the liquid-phase processes requires a special technique to separate them from the final product. An uncontrollable agglomeration of solid nanoparticles makes their industrial application impossible in a dry phase. Moreover, the toxic effects of nanoparticles on human health and ecosystem have been discussed in the literature.

Utilization of larger micron-size agglomerates of nanoparticles is one of the alternative ways to safely handle nanoparticles in the dry phase on the industrial scale. Spray drying is one of the well-known techniques that have been widely applied for production of microspheres from nano-sized particles. Using this technique, it is possible to control the size distributions as well as the morphology of spray-dried agglomerates. However, there are only a limited number of publications related to spray drying of LDH nanosuspensions. Moreover, spray-dried LDH agglomerates of rather wide size distribution, low surface area and irregular shape were reported in the literature.

In this study, we explore the effect of process parameters, such as spraying air flow rate, drying air temperature, slurry concentration and slurry feed rate on the size distribution, morphology, specific surface area and pore volume of LDH agglomerates fabricated by spray drying of nanosuspensions.

The round-shaped agglomerates with a diameter of more than 2 and less than 15 micrometers were obtained by spray drying of suspensions of nanoparticles with median diameter of 171 nm without addition of dispersant and binder. Our results confirmed that the spraying air flow rate has the dominant effect on the agglomerate median size and the width of agglomerate size distribution as well as on the agglomerate specific surface area. The small agglomerates with narrow size distribution were collected during drying at the high flow rate of spraying air. Our results also indicate that lowering the temperature of drying air and using the dilute nanosuspensions promote generation of small agglomerates.

The specific surface area of spray-dried LDH agglomerates is in the range of 49 to 101 m^2/g. The agglomerates of high specific surface area were obtained at low flow rate of spraying air, high slurry feed rate of concentrated slurry and using low temperature drying air. These results also confirm that drying at low drying rate yields to formation of agglomerates of high surface area. Similar trends were observed in terms of the variation of total pore volume with drying process parameters.

Thus, by adjusting the spray drying conditions, it is possible to produce agglomerates of LDH nanoparticles satisfying the product requirements.

513829 Impact of Spheroidization of UO_2 Powders on the Filling of Press Molds

Abstract:

In industry, the use of powders is very common. This underlines the interest in understanding the flow of powders in process lines (hopper, silos, transportâ€¦) and finding the link between powder physical characteristics and its rheological properties. This study deals with the flowing properties of a UO2 powder in a powder metallurgy process for nuclear fuel fabrication. A reproducible filling of the press molds is required to avoid mechanical grinding of the sintered pellets.

A powder is characterized by its particle size distribution but also by the morphology of particles. Morphology has a major impact on the cohesion and then on the flowability of powders [1]. UO2 powders are generally made up agglomerates which are composed of aggregates, themselves formed by crystallites (Figure 1). Agglomerates are primarily governing the powder flowing behavior. Their characteristics sometimes do not allow them to meet the expectations of the process, especially for the filling of the press molds.

For this reason, in a powder manufacturing process, a spheroidization operation (Figure 2) could precede the shaping of pellets to improve the filling properties.

In this study we observe the influence of particles morphology on the rheological properties of UO2 powders. Whereas there is no standard and universal method to characterize the flow of powders [2], there is direct (shear cells) and indirect methods (dynamic flowability measurements like permeability). We use both methods: the indirect methods based on qualitative measures of powder cohesion, the direct methods to determine the powder cohesion for a specific consolidation state through the Freeman shear cell [3]. While a shear cell measurement permits to determine the powder flow in quasi-static system, a permeability measurement shows the flow index in dynamic system. We identify the link between theses measurements and the quality of the filling.

References:
[1] Xiaowei Fu and others, «Effect of Particle Shape and Size on Flow Properties of Lactose Powders», Particuology, Advances in Characterization and Modeling of Particulate Processes, Apr. 2012.
[2] P. Juliano, B. Muhunthan, and G. V. Barbosa-Cánovas, « Flow and shear descriptors of preconsolidated food powders », J. Food Eng., vol. 72, no 2, p. 157-166, Jan. 2006.
[3] J. Schwedes and D. Schulze, « Measurement of flow properties of bulk solids », Powder Technol., vol. 61, no 1, p. 59-68, Apr. 1990.

508203 Modelling a Twin Screw Granulator Using the Discrete Element Method

Abstract:

Wet granulation is a process used to create larger stable agglomerates (granules) from fine powders. This has many desirable outcomes such as improving flowability, compactibility, and homogeneity. Granulation is commonly employed in the food, pharmaceutical, detergent, and fertilizer industries, but despite its wide adoption it is often inefficiently operated, with high recycle ratios in continuous processes and high rejection rates in batch processes. Wet granulation is the most common type of granulation and, in pharmaceutical applications, it is a critical step in tablet manufacturing that affects the uniformity and compactibility of the final dosage form.

Traditionally batch granulation was the favoured granulation process method in the pharmaceutical industry due to the challenges faced from adopting a continuous approach, such as cost or inability to monitor product quality reliably. Since the introduction of Quality by Design (QbD) and Process Analytical Technology (PAT) by the FDA in 2003 there has been a move towards new manufacturing techniques such as continuous processing.

Although tremendous efforts have been made to gain scientific insight into the granulating process, a fundamental understanding of wet granulation is still lacking due to the complexity of the mechanisms involved. With twin screw granulation becoming a popularly employed method of wet granulation, an in-depth understanding of particle enlargement in the granulating process is necessary in order to improve the quality of the final product without the need for large-scale Design-of-Experiment studies.

Despite the extensive experimental research carried out for twin screw granulators in recent years, there has been little computational work carried out. This paper employs the Discrete Element Method (DEM) to study a 25 mm diameter, GEA ConsiGmaâ„¢ 1 twin screw granulator with a typical 60Â° forward configuration for kneading elements. The DEM simulations were conducted using the commercial code EDEM with a DEM contact model developed for cohesive solids implemented through an API. The contact model is based on an elasto-plastic contact with adhesion and uses hysteretic loading and unloading paths to model the elastic-plastic contact deformation. In these simulations, the adhesion is used to capture the effect of the binder liquid without the complication of modelling the liquid directly. The adhesion parameter is a function of the plastic contact overlap. The model has previously been shown to be able to predict the stress-history-dependent behaviour depicted by a flow function of the material. In this study the effect of particle size, gap tolerance and cohesion on the residence time distribution and local solid fraction at different elements in the granulator is assessed. The level of cohesion used is related back to uniaxial test results which can be used to measure the flowability of a cohesive granular solid.

508216 Hybrid Multiscale Modelling of a Twin Screw Granulator

Abstract:

Wet granulation is a process used to create larger stable agglomerates (granules) from fine powders. This has many desirable outcomes such as improving flowability, compactibility, and homogeneity. Granulation is commonly employed in the food, pharmaceutical, detergent, and fertilizer industries, but despite its wide adoption it is often inefficiently operated, with high recycle ratios in continuous processes and high rejection rates in batch processes. Wet granulation is the most common type of granulation and, in pharmaceutical applications, it is a critical step in tablet manufacturing that affects the uniformity and compactibility of the final dosage form.

Modelling processes such as wet granulation is not straightforward, but may be achieved using techniques such as Population Balance Modelling (PBM) and Discrete Element Method (DEM). PBM groups particles into different classes based on their properties and uses empirically derived differential equations to describe key processes such as aggregation or breakage. However, although PBM tracks the size and number of particles over time, it cannot account for the distribution of the particles and the effect of this local variability.

By contrast, DEM simulations track individual particles, and as such, phenomena such as segregation or liquid migration can be studied. DEM simulations also consider the effect that the geometry and its dynamics have on the particles' behaviour, which is something that PBM cannot account for directly. However, DEM simulations of fine materials can be computationally expensive and this often results in reduced numbers of particles being used to make the simulation computationally tractable. This is usually achieved through increased, unrealistic particle sizes or the simulation of isolated local zones without considering adjacent zones which may affect the behaviour.

Traditionally, large-scale Design-of-Experiment (DoE) studies would be needed to estimate key rate parameters for use in PBM modelling of granulation and this can be both costly and time-consuming. DEM simulations provide an alternative method to estimate these parameters. By coupling the two simulation methods, one can utilise the key benefits of each method to better simulate complex processes.

This paper presents a hybrid approach to modelling twin screw granulation via DEM–PBM coupled simulations to study a 25 mm diameter, GEA ConsiGma™ 1 twin screw granulator with a typical 60° forward configuration for kneading elements. The DEM simulation was conducted using the commercial code EDEM with a DEM contact model developed for cohesive solids implemented through an API. The PBM simulation is carried out with the commercial code gFORMULATE which is built on top on the advanced modelling platform gPROMS. A bi-directional coupling is implemented with information shared between the two simulation tools at specified intervals. Preliminary results for this coupled approach are presented and discussed.

509577 Blocking Rules for Discharging Granular Materials from a Flat Bottom Hopper

Abstract:

Hoppers and silos are widely used in storing granular materials in various industries, such as agricultural, chemical, food and pharmaceutical industries. Three key aspects need to be considered when hoppers are used to store powders and grains: 1) stress analysis; 2) hopper blockage ; 3) mass flow rate. A hopper must be designed strong enough to sustain the stress generated from the bulk materials during storage and discharge. On the other hand, a hopper must be designed properly so that the stored bulk materials can be discharged, i.e. without blockage and formation of ratholes and arches. Furthermore, the mass flow rate needs to be predicted and controlled. In this study, a combined experimental and numerical study was performed to determine the blocking rule for discharging granular materials from a flat bottom hopper. Granular materials of various sizes were considered and the minimal orifice size through which the bulk materials can be discharged was determined experimentally using a Flodex tester and numerically using DEM. The dependency of the minimal

orifice size on particle size was explored. An semi-empirical model was also developed. Using this model, the critical orifice size at which the flow will stop, i.e. blockage takes place, can be determined from the mass flow rate data. It was found that excellent agreement between the numerical results and experimental measurements was obtained. Both experimental and numerical results showed that, even for spherical particles, layered dead zones could be formed during the discharge from the flat bottom hopper. In addition, mono-layered dead zones were developed when large particles were used. Furthermore, the developed blocking rules were also verified experimentally and numerically.

508310 Characterization of Mesoscopic Structure in Cohesive Powder, Neat or Blended, By X-Ray Computed Tomography and Prediction By the Discrete Element Method

Abstract:

Understanding and predicting structural characteristics of cohesive powder systems is a technical challenge in many powder technology processes and applications. The discrete element method (DEM) [1] is a promising approach to model the structure and flow behavior of cohesive granular systems. Advances in experimental techniques such as X-ray computed tomography (XRCT) have now enabled characterization of fine cohesive powder systems with sub-micron resolution. Combined with standard bulk powder characterization methods (e.g. density), this technique allows for mesoscopic structural characteristics of cohesive powder systems to be compared with DEM simulation. In this study, DEM and XRCT are used to characterize the mesoscale structure, under controlled consolidation conditions, of two moderately cohesive 'model' (roughly spherical) powders with distinct material properties. The DEM simulations are based on a constitutive elasto-plastic model [2] that makes use of the JKR theory [3] to account for cohesive interparticle forces. Density and void structure obtained from DEM simulation are compared to characterization results obtained from XRCT image analysis. XRCT and DEM simulation approach is applied to examine structures of binary mixtures of the two model powders. The ability of DEM simulation and the contact model to accurately predict structural characteristics of single component and binary granular systems is discussed.

References

[1] P. A. Cundall and Otto D.L. Strack. A discrete numerical model for granular assemblies. Geotechnique, 29(1):47–65, 1979.

[2] C. Thornton and Z. Ning. A theoretical model for the stick/bounce behavior of adhesive, elastic-plastic spheres. Powder technology, 99(2):154– 162, 1998.

[3] K.L. Johnson, K. Kendall, and A.D. Roberts. Surface energy and the contact of elastic solids. In Proceedings of the Royal Society of London A: Mathematical, Physical and Engineering Sciences, volume 324, pages 301–313. The Royal Society, 1971.

511638 Numerical Research of Hydrodynamics in Gas-Solid Micro Fluidized Beds

Abstract:

Due to the advantages of less energy consumption, more safety and good operability, the miniaturization of fluidized bed, or named micro fluidized bed, has been extensively studied by a lot of researchers. The Micro Fluidized Bed Reaction Analyzer (MFBRA) was adopted by the Institute of Process Engineering (IPE), Chinese Academy of Sciences (CAS), to study the gas-solid reaction kinetics under isothermal differential condition[1, 2]. In the MFBRA, the hydrodynamics of micro fluidized beds play a vital role on the heat and mass transfer and which is a key to the accurate measurement of gas-solid reaction kinetics. In order to optimize and design the MFBRA, numerous experiments have been conducted to study the hydrodynamics of micro fluidized beds[3, 4]. Unfortunately, those works mostly focused on the bed pressure drop and minimum fluidization velocity, nevertheless, the detail information, such as local solid volume fraction and particle velocity profile in micro fluidized beds are reported rarely. It is because the existing experimental techniques, such as, optical fiber and capacitance probes are hard to measure those local parameters without disturbing the flow field.

The objective of this work is to study the hydrodynamics of micro fluidized beds and to investigate the isothermal characteristics during kinetics testing of graphite powder combustion in MFBRA. In this study, Discrete Particle Model (DPM) and Two-Fluid Model (TFM) are selected to investigate micro fluidized beds, respectively. The DPM is carried out by the commercial software EDEM-Fluent Coupling simulations. And for TFM, we used the Eulerian multiphase model in the ANSYS Fluent. From our numerical simulations, the detailed hydrodynamic characteristics of micro fluidized beds are explored systematically by comparing different column sizes and superficial gas velocity. Meanwhile, it is found that both DPM and TFM have the ability to model the MFBRA under practical operation condition with high superficial gas velocity. In the term of isothermal characteristics in micro fluidized beds, the temperature field distribution as well as the heat transfer between heat-carrier and graphite powder are explored by the combination of simulations and experiments.

Reference:

1. Xu, G., *Micro fluidized bed reaction kinetics analyzer. Supported by instru-ment development project of Chinese Academy of Sciences, China, Project numberY2005014. (in Chinese).* 2005.
2. Yu, J., et al., *Isothermal differential characteristics of gasâ€"solid reaction in micro-fluidized bed reactor.* Fuel, 2013. **103**: p. 29-36.
3. Liu, X., G. Xu, and S. Gao, *Micro fluidized beds: Wall effect and operability.* Chemical Engineering Journal, 2008. **137**(2): p. 302-307.
4. Guo, Q.j., Y. Xu, and X. Yue, *Fluidization Characteristics in Micro-Fluidized Beds of Various Inner Diameters.* Chemical Engineering & Technology, 2009. **32**(12): p. 1992-1999.

507312 An Improved Bubble-Based Drag ModelÂ for Accurate Coarse-Grid Two-Fluid Modeling of Geldart a Powder Bubbling Fluidization

Abstract:

An improved bubble-based energy minimization multi-scale (EMMS) drag model was formulated in which the original bubble-based EMMS drag model at macroscale bed level was extended to microscale sub-grid level for accurate coarse-grid two-fluid modeling (TFM) of Geldart A powder bubbling fluidization. At both the macroscale and microscale levels, the physically heterogeneous structures were modeled by three pseudo phases, i.e., bubble phase, emulsion phase, and inter-phase, where conservation and stability condition equations were applied. Through such scale and phase decompositions, the improved bubble-based EMMS drag model was formulated through a two-step macro-to-micro scheme, where the obtained drag coefficient is related to phase volume fractions, velocities, and accelerations within each coarse grid. Modeling results show that TFM combined with the improved bubble-based EMMS drag model can accurately reproduce the heterogeneous structures for a laboratory-scale Geldart A powder bubbling fluidized bed. At relatively coarse grid, the predicted axial and radial distributions of solid concentration agree very well with experimental measurements. This indicates that our proposed improved bubble-based EMMS drag model incorporated TFM is promising for accurate and economical simulation of engineering-scale Geldart A powder bubbling fluidization.

507860 Numerical Simulation on Fine Particle Transport Behaviour in Electrostatic Precipitators

Abstract:

Electrostatic precipitators (ESPs) are a major air pollution control device to remove particles from treated gas. Extensive research has been done on several aspects such as electrostatics, fluid dynamics, charging mechanism, and particle dynamics. The overall particle-collection efficiency of a modern industrial wire-plate ESP exceeds 99%. However, the removal efficiency of submicron particles by conventional ESPs is as low as 70â€"80% owing to the low charge carried by fine particles and the increase in mobility with decreasing size. Submicron particles possess a high superficial area, thus favouring toxic heavy metals and other pollutants. Therefore, fine particles are collected with much lower efficiency and further research is required to improve this aspect.

In this work, a numerical study on the charging and transport behaviour of fine particles has been carried out based on wire-plate electrostatic precipitators (ESPs) with multiple wire electrodes. The effect of the applied wire voltage, inlet height, and wire distance on particle charging and transport behaviour, and the influence of the precipitator structure on particle trapping are analysed in detail. The 2-D numerical model involves electric field, space-charge density, gas flow, and particle trajectories. The fluid-flow velocity is higher than 1 m/s in this study and the EHD flow has a negligible. A Lagrangian approach for particle transport is utilized with the trajectories of individual particles tracked by solving their equation of motion. The electric field and gas flow are calculated using the finite-volume method. The flow simulation is done using the commercial software FLUENT with other parts coded as user-defined functions. The following main results can be drawn:

(1) It was observed that the Y-displacement of the particles exhibited a universal increase for all particle sizes, especially for larger particles, at low injection positions (Y < 25 mm). The larger the particle size, the higher the Y-displacement of particles because larger particle sizes improve both the particle charging level and electric field strength. No significant improvements were found for submicron particles.

(2) Increasing the spacing of the discharge electrode within a certain range and therefore increasing the electric field, it was observed that the particle-trapping effect improved and the outlet level exhibited a universal increase for all particle sizes, especially for large particles.

(3) The ESP models M1 and M2 were used to investigate their influence on the outlet height. Except for particles with a diameter of 2 μm, all other particle sizes showed an increased outlet height in M2 compared to that in M0. In addition, the different discharge-electrode arrangements, especially in M2, exhibited significant effects on the charging and transport behaviour of particles with a diameter of 1 μm.

(4) Particles with an inlet height of 5 and 10 mm were studied in M1 and M2 precipitators. The changed discharge-electrode arrangement improves particle trapping, especially in M2. This is because of the electron trajectories being closer to the discharge electrodes (M2), giving the particles more charge and leading to a greater electric field.

507879 Lattice Boltzmann Simulations of Porous Particulate Flows

Abstract:

The comprehensive understanding of the particle-fluid interaction is very important for fluid-particles system, and thus many studies have been conducted for the particle-fluid system[1-3] of solid impermeable particles. In the real industrial fields such as the catalyst process in chemical engineering[4], the particles usually possess the porous structure which will have effects on the fluid-particles interaction[5-6]. However, there has been little open literature discussing the particle-fluid system of porous permeable particles. Therefore, the objective of the present study is to investigate the effect of the porosity and permeability on the porous particulate flows which are formulated by the volume-averaged governing equations[7] in terms of intrinsic phase average velocity. The lattice Boltzmann equation model (LBE)[8] is used to solve the governing equations. Firstly, we validate the LBE model by the numerical results existing in the literature for the flow around and through one porous particle. Then, the porous particles settling against gravity in a quiescent fluid, i.e., the DKT process[9] for porous particles, is investigated. The effects of the porosity and Darcy number on the porous particulate flows are studied numerically in details. Also, we present the simulation results of the fluidization of 512 porous particles. Our results demonstrate the significance of the porous structure of particles for the particles dynamics behaviors.

Reference

[1] Ladd, A.J.C., 1994. Numerical simulations of particulate suspensions via a discretized Boltzmann equation Part I. Theoretical foundation. J. Fluid Mech., 271, 285-310.

[2] Hill, R.J., Koch, D.L., Ladd, A.J.C., 2001. The first effects of fluid inertia on flows in ordered and random arrays of spheres. J. Fluid Mech. 448, 213-241.

[3] Van der Hoef, M.A., Beetstra, R., Kuipers, J.A.M., 2005. Lattice Boltzmann simulations of low-Reynolds-number flow past mono- and bidisperse arrays of spheres: results for the permeability and drag force. J. Fluid Mech. 528, 233-254.

[4] Zhao, Y.F., Li, H., Ye, M., Liu, Z.M., 2013. 3D numerical simulation of a large scale MTO fluidized bed reactor. Ind. Eng. Chem. Res. 52, 11354-11364.

[5] Noymer, P.D., Glicksman, L.R., Devendran, A., 1998. Drag on a permeable cylinder in steady flow at moderate Reynolds numbers. Chem. Eng. Sci. 53, 2859-2869.

[6] Zhu, Q.Y., Chen, Y.Q., Yu, H.Z., 2014. Numerical simulation of the flow around and through a hygroscopic porous circular cylinder. Comput. Fluids 92, 188-198.

[7] Wang, L., Wang, L.-P., Guo, Z.L., Mi, J.C., 2015. Volume-averaged macroscopic equation for fluid flow in moving porous media. Int. J. Heat Mass Trans. 82, 357-368.

[8] Chen, S.Y., Doolen, G.D., 1998. Lattice Boltzmann method for fluid flows. Annu. Rev. Fluid Mech. 30, 329-364.

[9] Fortes, A.F., Joseph, D.D., Lundgren, T.S., 1987. Nonlinear mechanics of fluidization of beds of spherical particles. J. Fluid Mech. 177, 467-483.

508370 Modeling of a Novel Multi-Particle Collision Model for Gas-Solid Flows

Abstract:

Fluidized beds are widely applied in the chemical, petrochemical, metallurgical, environmental and energy industries due to their favourable mass and heat transfer characteristics. In past few decades computational fluid dynamics has proven to predict hydrodynamics of fluidized beds. DPM (Euler-Lagrange) [1] and TFM (Euler-Euler) [2] are two most common modelling approaches for gas-solid flow in fluidized beds. In TFM the solid phase itself is treated as a continuum phase interpenetrating with the gas phase. Since the concept of individual particles has disappeared, the effects of particle-particle interactions in the TFM has to be incorporated in a statistically averaged way, via the solid phase pressure and viscosity. TFM is preferred model to study large scale fluidized beds but model suffer from high numerical diffusion. And the KTGF theory in TFM assume elastic collisions and is limited only to spherical particles [2]. In DPM, the gas phase is considered as the continuous phase whereas the particle trajectories are obtained by integrating Newtonâ€™s equations of motion. Particle-particle and particle-wall interactions are explicitly taken into account using various physical models such as the soft sphere model. DPM proves to be accurate modelling of gas-particle flow. However, lagrangian models quickly become computationally too expensive because of the large number of particle interactions. Therefore coarse graining of discrete particles is an emerging model that can handle large number of particles effectively. The new coarse-grained model developments are of vital importance for the advancement of large-scale modelling of gas-particle flows that are flexible to extended for polydispersity, and non-spherical particles. In this study a novel lagrangian multi-particle collision (MPC) method is developed for predicting dense particulate-fluid flow.

The MPC solver is developed in OpenFoam [5], where the fluid phase is treated as a continuum and the solids are modelled by lagrangian particles. The MPC model grew out of the multiphase particle-in-cell (MP-PIC) model [3], that employs a fixed Eulerian grid, and particles are traced in a lagrangian fashion, including a parcel approach where each simulated particle represents a large collection of real particles. Each parcel follows Newtonâ€™s equations of motion, where inter-phase momentum transfer is accounted through mapping and interpolating quantities back and forth between parcel locations and the eulerian grid. Particle collisions in MP-PIC are not fully resolved but are derived from particles stresses that are theoretically calculated from Harris and Crighton [4] stress closure. In the MPC model, we implemented a functional form for the particle stress tensor where the coefficients are obtained computationally by means of DPM simulations. The idea is to extend the model from spherical particles to non-spherical particles, for which theoretical particles closures are currently not available in the literature. The viscous stress and random forces are typically ignored in MP-PIC model due to their insignificant effect [3]. Importantly, we incorporated viscous stress and random forces representing scattering of particles by collisions, which are obtained from DPM simulations by measuring the actual variance. Details of model development and validation is presented in this study.

Results from MPC model are compared with MPPICFoam and, with DPM simulations from DEM-LIGGGHT [6], and MFiX [7] code respectively. Pseuduo-2D fluidized beds of dimension 0.01x0.15x0.45 m is considered for comparison with experiments [8]. Comparison is made on the basic of pressure drop, void fraction, bubble dynamics and solids motion. We observed that MPC is in good agreement with DPM, and experiments, when compared to MPPICFoam. Unrealistic over-packing of solids is predicted in MPPIC due to limitation of Harris and Crighton [4] model and the velocity field updates, whereas in MPC these limitations overcomes and no over packing of particles is predicted. Solid circulations are in good agreement with DPM results, where solid moving above in the center and downward near to the wall. Time-average porosity plots revels the path of bubble motion and is in close agreement with results of DPM. In conclusion a novel multi-particle collision model is developed and validated to study gas-solid fluidized beds. The model is efficient to handle large number of particles. In future study, we plan to extend the model for dense flow of non-spherical particles.

Acknowledgments

The authors thank the European Research Council for its financial support under its consolidator grant scheme, contract no. 615906 (NonSphereFlow).

References

1. N.A. Patankar, D.D. Joseph, 2001. Modeling and numerical simulation of particulate flows by the Eulerian Lagrangian approach, International Journal of Multiphase Flow, Volume 27, Issue 10, October 2001, Pages 1659-1684.
2. Gidaspow, D., 1994. Multiphase flow and fluidization continuum and kinetic theory description, Academic Press, Boston.
3. Andrews, M.J., O'Rourke, P.J., 1996. The multiphase particle-in-cell (MP-PIC) method for dense particulate ows, Int. J. Multiphase Flow 22, 379-402.
4. Harris, S.E., Crighton, D.G., 1988. Solutions, solitary waves and voidage disturbances in gas-fluidized beds, J. Fluid Mech. 195, 437.
5. OpenFOAM, 2015. The Open Source CFD Toolbox, Ver. 3.0.0. ESI Group, Paris, France.
6. https://www.cfdem.com/liggghtsr-open-source-discrete-element-method-particle-simulation-code
7. https://mfix.netl.doe.gov/mfix/
8. M. Goldschmidt, 2001. PhD thesis, Hydrodynamic Modelling of Fluidised Bed Spray Granulation.

508490 Dpm Analysis of Large Fluidized Catalytic Cracking (FCC) Reactors

Abstract:

DPM Analysis of Large Fluidized Catalytic Cracking (FCC) Reactors

Abstract:

Computational Fluid Dynamic (CFD) modeling is a powerful tool for equipment design and development and is widely used at UOP, particularly for the reactor and regenerator design of FCC process unit. Through CFD modeling, a technical solution was developed for a large capacity FCC process unit purposing an elevated amount of propylene production.

A key challenge in FCC process design is how to optimize the size of the reactor vessel, which is a function of the number and size of reactor cyclones (catalyst separationdisengagers). UOP developed a new cyclone cluster configuration via CFD study of the reactor internals, including UOP's VSS™ riser termination device and reactor cyclones, which reduce the reactor size significantly compare to the conventional design, while maintaining effective catalyst / vapor separation and mechanical reliability. The objective of the current CFD study is to evaluate the catalyst separation efficiency and performance of the new riser termination device technology, which accommodate more cyclones in a smaller reactor vessel.

To better understand the flow pattern, flow distribution and the gas residence time inside the UOP VSS™ , the analysis was broken down into two sections. Part 1 focused on single gas phase only and in part 2 the particles behavior is studied using the Discrete Phase Model (DPM) in Fluent.

The single gas flow analysis results showed that the flow in the VSS chamber section has the same rotational pattern as in conventional FCC/VSS units, where the catalyst will separate from the gas products. Also the gas flow distribution at each chamber arm is uniform, which is critical to catalyst separation efficiency.

The DPM analysis confirmed an equal catalyst particle distribution through all the cyclones and a high efficiency separation in the riser termination device. Having this information from CFD allows for the new design to move forward.

509623 Catalytic Reactor Design Using Multiphase CFD

Abstract:

Catalytic Reactor Design Usng Multiphase Cfddimitri Gidaspow the 2011 Book , "Computational Gas-Solids Flows and Reacting Systems ", Edited By S. Pannala, M. Syamlal and Tom O'Brien ,Engineering Science Reference, Shows the Progress We Have Made in Using Multiphase CFD in Modeling Fluidized Bed Chemical Reactors and Other Multiphase Systems, Such As the Pyroplastic Flow Emitted from Volcanoes. in This Presentation the Design of Two Types of Catalytic Reactors Will be Described: A Fischer_ Tropsch Slurry Bubble Column Reactora Polymerization Reactor the CFD Design of a New Slurry Bubble Column Reactor Was Already Described By Gidaspow ,Et Al in Chemical Engineering Science 134,784-799 (2015).in This Presentation the Design Will be Extended to Obtain 100, 000 Barrels of Diesel per Day, Twice That of the Largest Commercial Reactors. the Production of Synthesis Gas Using Oxygen or Carbon Dioxide in Catalytic Fluidized Beds Needed for the Fischer-Tropsch Reactor Will Also be Discussed Briefly. for the Polymerization Reactor Our CFD Code Described in " Computational Techniques " By D. Gidaspow and V. Jiradilok , Nova Science, 2009 Was Modified for Polymerization Reactors. Such Reactors and Their CFD Modeling Are Described in a Chapter in Pannala's Et Al Book from Professor Rodney Fox's Group. They Point out That the Fine Particle (80 microns) Catalyst Radial Distribution, Is a Remaining Issue the Computations to be Presented Will Show That the Catalyst Will Have a High Concentration at the Reactor Walls for Various Operating Conditions. This Leads to a High Rate of Reaction at the Reactor Walls and May Result in Melting of the Polymers and Sheet Formation , Forcing Reactor Shut-Down. in Some of Our Simulations There Is Large Bubble and Slug Formation Similar to Those Mentioned in Fox's Chapter. Acknowledgement : I Thank Dr. Mayank Kashyap of SABIC for a Valuable Discussion and His Papers " Application of Multiphase Flow CFD in the Gas Phase Polymerization Process " Fluidization XV. May 24,2016 and a Related Paper Presented at Doe NETL, July 2016

509627 Simulation of Large Particle Turbulent Fluidization in Riser Reactors By Coarse Grain DEM-CFD

Abstract:

The complex two-phase flow in the riser is an important element of many industrial circulating fluidized bed reactors. In some cases, e.g. gas-solid polymerization, relatively large particles are transported upward in a turbulent regime, in which the high gas velocity results mostly from heat removal requirements. Strong solids acceleration, radial and axial velocity profiles for both the gas and solid, intense clustering and inhomogeneities make it impossible to reliably predict the riser head or local bed density profile. Detailed numerical simulations proved capable of showing some of the peculiarities of the flow, but limitations in the scale, lack of grid independence, need for special drag corrections and sub-grid models still hinder confident use of such tools in design and scale-up, particularly in industrial applications characterized by particle size and Reynolds numbers that are not well covered by the available scientific literature.

In the present contribution, coupled Discrete Element Method-Computational Fluid Dynamics (DEM-CFD) simulations are carried out using a coarse graining technique for the solid particles. According to this method, small number of particles are grouped into a computational particle (*parcel*) and represented by it as a spherical "lump," to reduce the computational cost of the simulations and allow lab-scale systems to be represented in full detail. The investigation focuses on the influence of the coars-graining degree, i.e. the number of real particles represented by a *parcel*, on the simulated flow hydrodynamics both in terms of its advantages (a quantification of the computational savings) and disadvantage, primarily the real capability of the method to resolve the behaviour observed and produce results in line with experimental findings. The gas and solid motion are studied, with the aim to characterize the clustering phenomena observed during the up-flow and its effect on the pressure drop. Assuming prescribed gas and solid inlet velocity fields, the potential tendency to show core-annulus flow, the intensity of solids clustering and the local bed density within the riser are examined at different number of particles per *parcel*.

509640 Numerical Study of Particles Shape Effects on Solid-Liquid Fluidizations

Abstract:

Numerical study of particles shape effects on solid-liquid fluidizations

Esmaeil Abbaszadeh Molaei, Zongyan Zhou*

Laboratory for Simulation and Modelling of Particulate Systems, Department of Chemical Engineering, Monash University, Clayton, VIC 3800, Australia

Solid-liquid fluidizations have been observed in many industries due to their high heat and mass transfer rate, high operation flexibilities, and reduced back mixing of fluid and solid phases. In most of these multiphase operations, particle properties such as size, density, and shape may experience significant changes during the process because of attrition, coalescence, comminution or chemical reactions. The difference in particle properties can result in significant segregation problem in liquid fluidized beds. To gain an understanding of solid-liquid fluidisation behaviour, most existing studies, either experimentally or numerically, have focused on solid flow behaviour with mono-sized particles or binary and ternary mixtures with different particle sizes or densities. In spite of substantial studies on the hydrodynamics of liquid fluidized beds, the effects of particles shape on the solid-liquid flow behaviour are still less reported. Except for two physical experiments, the previous studies considered particles as spheres, and few studies have been remarked on liquid fluidization of non-spherical particles. However, particles are generally irregular in most of the processes such as ore beneficiations using liquid-solid fluidized beds separator. Particle shape can cause different flow behaviour in liquid fluidizations rather than that observed for ideal spherical particles. Nonetheless, the answers to some fundamental questions such as how and why particle shape affects flow phenomena are not clear, and hence still poorly understood.

Therefore, in this work, a combination of computational fluid dynamics (CFD) and discrete element model (DEM) [CFD-DEM] is used to examine the effects of particle shape on the flow characteristics in liquid fluidizations. In the simulations, ellipsoidal particles are used as they can represent a wide range of particle shapes from oblate to prolate particles. Different aspect ratios are employed, and results are analysed mainly in terms of particle flow patterns, pressure drop, and bed expansion. In addition, force and particle orientation analysis is also conducted to examine the effect of particle shape and explain the occurrence of entrainment phenomenon.

The results show that ellipsoidal particles intend to enter to freeboard region and entrainment phenomenon occurs, and consequently, the interface between the bed and freeboard cannot be observed clearly. In addition, the pressure gradient variation with bed height is uniform for spheres; however, this uniformity is replaced with â€œdistorted parentheses-shapeâ€ profile for disc-shape and elongated particles. Finally, bed expansion analyses show that spherical particles have the lowest bed expansion. On the other hand, disc-shape particles intend to expand more significantly than spherical and elongated particles. This causes the entrainment phenomenon which has been found to be due to the effects of particles shape on particles terminal velocity.

* Corresponding author. Tel: 61 3 9905 0846; email: zongyan.zhou@monash.edu.

514318 CFD Investigation on Gas-Solids Flow and Heat Transfer in Two Fluidized Catalyst Cooler

Abstract:

Our previously experimental work showed that the internal circulation of solids in an annular catalyst cooler (ACC) had an ability to enhance the heat transfer between bed and wall, comparing with the base catalyst cooler (BCC). The current work simulated in detailed hydrodynamics of gas-solids flow and heat transfer to advance our understanding of the ACC by using the two-fluid model. The effective particle diameter $d_p^{\text{âŽ}}$ was determined to 170 Î¼m and the specularity coefficient Ï† was 0.3 in the 3D simulations. The catalyst cooler can be divided to four distinct regions including draft tube region, annulus region, bottom region and particle diffluence region according to the distributions of solids volume fraction and solids velocity. The effect of gas-solids flow on bed-to-wall heat transfer was compared in the different regions of the BCC and ACC. And the effects of operational gas velocity and bed structure were discussed. The results showed that the proper selection of the ratio of gas velocities and the position of the heat transfer tube were needed to form a stable internal solids circulation in the ACC. The ACC had a combined hydrodynamic feature of up-flow and down-flow catalyst coolers with bigger solids volume fraction and smaller particle resident time, which are beneficial for improving the heat transfer between solids and wall.

Hydrodynamics and mixing characteristics of a new-type particle mixer

Mengxi Liu, Chunxi Lu, Zhenliang Meng

State Key Laboratory of Heavy Oil Processing, China University of Petroleum, Beijing 102249, PR China

E-mail: mengxiliu@sina.com (M. Liu)

Abstract:

In order to further enhance the particle mixing in the internally circulating fluidized bed, especially along the radial direction, a new-type particle mixer with slots on the draft tube was proposed in the present work. The hydrodynamics of the mixer was investigated by using a multi-scale computational fluid dynamics with the EMMS drag. The prediction in terms of cross-sectionally averaged solid holdup and solid velocity were compared to experimental data for validation. The results show that in draft tube a low pressure region can be observed near the slots and in annulus the pressure in the vicinity of the slots seems larger than that in the region between the adjacent slots, which induces horizontal particle immigration and mixing. Nearly 25.4% to 29.7% of gas in the draft tube bypasses and flows through the slots into the annulus. The ratio of the mass flux of the tracer gas bypassing into the annulus to that of total tracer gas is approximately equal to 0. With increase of the superficial gas velocity in the draft tube from 0.3 m/s to 0.45 m/s, the total solid mass flux increases from 4.87 kg/s to 10.86 kg/s, while the mass flux through slots decreases from 2.69 kg/s to 0.52 kg/s.

1. Introduction

The characteristics of particle flow and mixing have a considerable influence on the mass and heat transfer, as well as the product quality. According to the previous study, the mixing degree of particles along the axial and radial direction is quite different [1-5]. Mostoufi and Chaouki [4] discovered that the diffusivities varied from 3.3×10^{-3} m^2/s to 5.6×10^{-2} m^2/s along the axial direction and 2.6×10^{-4} m^2/s to 1.5×10^{-3} m^2/s along the radial direction. Berruti et al. [6] also revealed that the particle mixing rate for radial direction was an order of magnitude smaller than for the axial direction. Therefore, the weak particle flow along the radial direction greatly limits the mixing ability of a fluidized bed. Internally circulating fluidized bed (ICFB) has been widely recognized as an efficient fluidized bed reactor, in which orderly and macroscopic particle circulation occurs. Compared with the conventional fluidized bed, ICFB offers multiple particle circulation between the draft tube and the annulus, which leads to relatively strong particle convection, diffusion and shear, and thereby improves the particle mixing performance along the radial direction [7]. In order to further enhance the particle mixing for some processes in which a uniform particle mixing is required, a new-type particle mixer with slots on the draft tube was proposed in the present work.

With increase of the computational power, the computational fluid dynamics (CFD) provides an additional tool to predict particle flow and mixing in the local region, which is difficult with the experimental methods. In particular, the gas-solid flow in a gas-solid fluidized bed can be accurately described by the energy minimization multi-scale (EMMS) model [8]. In this paper, the hydrodynamics of the mixer was investigated by using a multi-scale CFD based on the EMMS drag model [9, 10].

2. Experimental apparatus, Materials and Operating conditions

The structure of the particle mixer is shown in Fig.1. The mixer has a height of 5 m and an inner diameter of 286 mm. A draft tube with a height of 1 m and an inner diameter of 215 mm is coaxially mounted into the column. The four rectangular slots of 80×40 mm are symmetrically opened on the draft tube at a height of h=860 mm, in order to strengthen the particle radial flow. A gas distributor plate and a ring distributor are installed below the draft tube and the annulus respectively. The height of the gap between the gas distributor plate and the draft tube is 60 mm. For the sake of the convenience, the mixer is divided into six zones: the pre-mixing zone, feed influence zone, draft tube zone, gas-solid separation zone, annulus zone and the bottom zone. A perforated plate is mounted at the bottom of the pre-mixing zone, which is connected with a central downcomer, extending into the draft tube. Particles are first mixed in the pre-mixing zone and then the pre-mixed particles flow downward through the central downcomer and into the feed influence zone. In the feed influence zone, the pre-mixed particles are mixed with the particles moving upward in the draft tube, flow upward and circulate between the draft tube and the annulus. In the annulus, particles move downward to the bottom. Then a part of particles flow out of the mixer and the other part flow horizontally into the draft tube and recirculate.

The solid particles used in the experiment were FCC catalyst (ρ_p=1498 kg/m^3), with an average diameter of 76 μm. Air (ρ_g=1.225 kg/m^3) was introduced into the mixer to fluidize the particles. The superficial gas velocity ranged from 0.3 to 0.45 m/s in the draft tube (based on the cross-sectional area of the draft tube) and remained at 0.05m/s in the annulus (based on the cross-sectional area of the annulus). The mass flux of cold particles remained at 13.8 kg/m^{2}s. As we only investigated the influence of slots on the flow and mixing of particles in this paper, no hot particles were induced into the mixer.

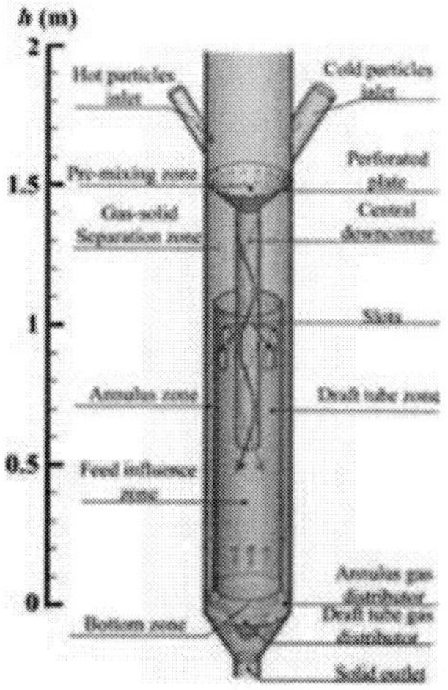

Fig. 1 Schematic diagram of the new-type particle mixer

3. CFD model and simulation method

The hydrodynamics of the mixer was performed using the Eulerian- Eulerian of Ansys Fluent

®13.0. The governing equations [11] include a set of mass and momentum equations for each phase, as shown in Table 1.

A 3-D mesh containing 580,000 cells was generated by using Gambit®2.4. The no-slip boundary was assigned for the gas phase and the partial-slip boundary condition developed by Johnson and Jackson [12] was used for the solid phase. In order to ensure a steady-state operation of the mixer, simulations were first performed for 30 s, and then the time-averaged data was analyzed for another period of 10 s. Other parameters settings in the model are shown in Table 2.

Table 1 Governing equations

Gas-phase Continuity equation $\frac{\partial}{\partial t}(\varepsilon_g \rho_g) + \nabla \cdot (\varepsilon_g \rho_g \boldsymbol{u}_g) = 0$

Solid-phase Continuity equation $\frac{\partial}{\partial t}(\varepsilon_s \rho_s) + \nabla \cdot (\varepsilon_s \rho_s \boldsymbol{u}_s) = 0$

Gas-phase momentum equation $\frac{\partial}{\partial t}(\varepsilon_g \rho_g \boldsymbol{u}_g) + \nabla \cdot (\varepsilon_g \rho_g \boldsymbol{u}_g \boldsymbol{u}_g) = -\varepsilon_g \nabla p + \nabla \cdot \boldsymbol{\tau}_g + \varepsilon_g \rho_g \boldsymbol{g} - \beta(\boldsymbol{u}_g - \boldsymbol{u}_s)$

Solid-phase momentum equation $\frac{\partial}{\partial t}(\varepsilon_s \rho_s \boldsymbol{u}_s) + \nabla \cdot (\varepsilon_s \rho_s \boldsymbol{u}_s \boldsymbol{u}_s) = -\varepsilon_s \nabla p - \nabla p_s + \nabla \cdot \boldsymbol{\tau}_s + \varepsilon_s \rho_s \boldsymbol{g} + \beta(\boldsymbol{u}_g - \boldsymbol{u}_s)$

Granular temperature equation

$$\frac{3}{2}[\frac{\partial}{\partial t}(\varepsilon_s \rho_s \Theta_s) + \nabla \cdot (\varepsilon_s \rho_s \Theta_s \boldsymbol{u}_s)] = (-p_s \boldsymbol{I} + \boldsymbol{\tau}_s) : \nabla \boldsymbol{u}_s + \nabla \cdot (\kappa_s \nabla \Theta_s) - \gamma_s - 3\beta\Theta_s$$

Gas-phase stress $\boldsymbol{\tau}_g = \varepsilon_g \mu_g [\nabla \boldsymbol{u}_g + (\nabla \boldsymbol{u}_g)^T] - \frac{2}{3} \varepsilon_g \mu_g (\nabla \cdot \boldsymbol{u}_g) \boldsymbol{I}$

Solid-phase stress $\boldsymbol{\tau}_s = \varepsilon_s \mu_s [\nabla \boldsymbol{u}_s + (\nabla \boldsymbol{u}_s)^T] + \varepsilon_s (\lambda_s - \frac{2}{3} \mu_s)(\nabla \cdot \boldsymbol{u}_k) \boldsymbol{I}$

Solid-phase pressure $p_s = \varepsilon_s \rho_s \Theta_s [1 + 2(1 + e_s) \varepsilon_s g_0]$

Solid-phase shear viscosity $\mu_s = \mu_{s,col} + \mu_{s,kin} + \mu_{s,fr}$

Where $\mu_{s,col} = \frac{4}{5} \varepsilon_s \rho_s d_p g_0 (1 + e_s) \sqrt{\Theta_s / \pi}$, $\mu_{s,kin} = \frac{10 \rho_s d_p \sqrt{\Theta_s / \pi}}{96 \varepsilon_s (1 + e_s) g_0} [1 + \frac{4}{5} g_0 \varepsilon_s (1 + e_s)]^2$, $\mu_{s,fr} = \frac{p_s \sin\phi}{2\sqrt{I_{2D}}}$

Radial distribution function $g_0 = [1 - (\varepsilon_s / \varepsilon_{s,max})^{1/3}]^{-1}$

Diffusion coefficient $\kappa_s = \frac{150 \rho_s d_p \sqrt{\Theta_s / \pi}}{384(1 + e_s) g_0} [1 + \frac{6}{5} g_0 \varepsilon_s (1 + e_s)]^2 + 2 \rho_s \varepsilon_s^2 d_p (1 + e_s) g_0 \sqrt{\Theta_s / \pi}$

Collisional energy dissipation $\gamma_s = \frac{12(1 - e_s^2) \varepsilon_s^2 \rho_s g_0 \Theta_s^{3/2}}{d_p \sqrt{\pi}}$

Heterogeneity index $H_d = \frac{\beta_e}{\beta}$, $\beta_e = \beta H_d$

Drag based homogeneous structure
$$\beta = \begin{cases} \frac{3}{4} C_d \frac{\varepsilon_s \varepsilon_g \rho_g |\boldsymbol{u}_g - \boldsymbol{u}_s|}{d_p} \varepsilon_g^{-2.65}, \varepsilon_g > 0.8 \\ 150 \frac{\varepsilon_s^2}{\varepsilon_g} \frac{\mu_g}{d_p^2} + 1.75 \varepsilon_s \frac{\rho_p}{d_p} |\boldsymbol{u}_g - \boldsymbol{u}_s|, \varepsilon_g \le 0.8 \end{cases}$$

Where
$$C_d = \begin{cases} 0.44 & \mathrm{Re}_s \ge 1000 \\ \frac{24}{\mathrm{Re}_s}(1 + 0.15 \mathrm{Re}_s^{0.687}), & \mathrm{Re}_s < 1000 \end{cases}, \quad \mathrm{Re}_s = \frac{\varepsilon_g \rho_g d_p |\boldsymbol{u}_g - \boldsymbol{u}_s|}{\mu_g}$$

Table 2 Simulation Parameters

Gas-solid model	Eulerian-Eulerian
Solids Pressure	Lun et al.
Particle-wall specularity coefficient	0.5
Gas-wall friction coefficient	No-slip
Pressure-Velocity Coupling	Simple
Gradient discretization	Least squares cell based
Momentum discretization	Second-order upwind
Volume fraction discretization	Quick
Max. number of iterations per time step	40
Pressure outlet (atm)	1
Inventory (kg)	56
Incipient solid holdup	0.6

4. Results and discussion

4.1 Model validation

Fig. 2 gives the distribution of cross-sectionally averaged solid holdup and particle velocity in the mixer. As shown in Fig. 2(a), the solid holdup is larger in the bottom zone (h=0.04 m) and due to the accelerated upward motion of particles, the solid holdup decreases with increase of the axial height (when $h \leq 0.16$ m). The cross-sectionally averaged solid holdup is nearly constant in the draft tube and gas-solid separation zone of $0.4 \leq h \leq 1.25$ m. The cross-sectionally averaged particle velocity is shown in Fig. 2(b). The particle velocity is smaller in the bottom zone and keeps almost invariable in the draft tube and the gas-solid separation zone. Compared with the experimental data, the particle velocity seems slightly higher, while the solid holdup in the simulation agrees reasonably with the experimental results.

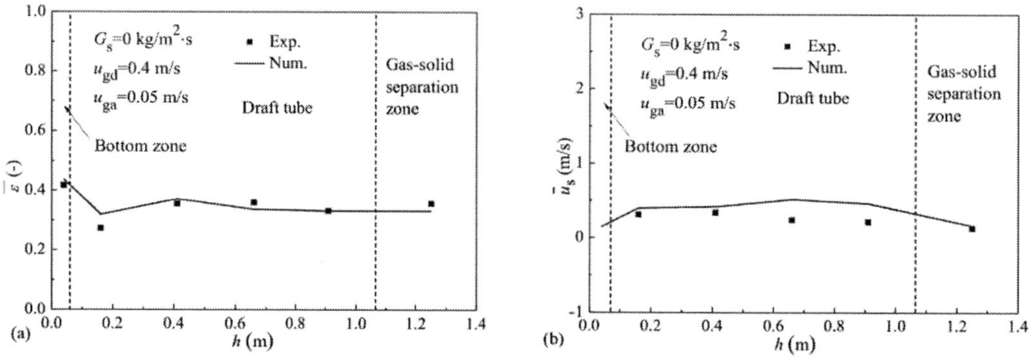

Fig. 2 Distribution of cross-sectionally averaged solid holdup and particle velocity as a function of axial heights

4.2 Pressure distribution near the slots

Fig. 3 shows the time-averaged contours of static pressure on the cross-section of h=0.9 m, the cross-section on the middle of the slots. It is seen that in the draft tube a low pressure region can be observed near the slots, smaller than the pressure in the region between the adjacent slots. This pressure drop forced particles inside the draft tube to move towards the slots and pass through them, which significantly facilitates the mixing of the particles along the radial direction. In the annulus the pressure in the vicinity of the slots seems larger than that in the region between the adjacent slots, inducing horizontal particle immigration and mixing.

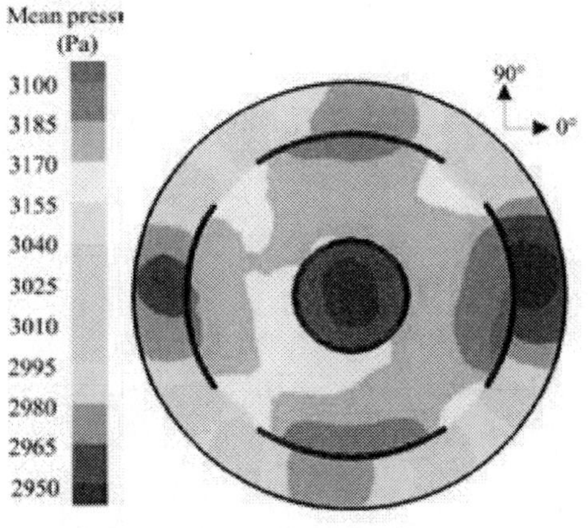

Fig. 3 Time-averaged contours of static pressure at the cross-section of the slots
(u_{gd} =0.4 m/s, G_s=13.8 kg/m^{2}s, h=0.9 m)

4.3 Gas bypassing

In the present work, the gas bypassing behavior was investigated by CFD simulation, and a tracer gas having identical physical properties with air was defined. After the mixer running stably, the tracer gas was introduced into the mixer from the gas distributor plate to replace the original gas. A cross section (h=0.06 m) on the exit of the annulus was selected and the mass flux of the tracer gas through the cross section was registered. The results are shown in Fig. 4. f_B refers to the ratio of the mass flux of the tracer gas through the cross section (h=0.06 m) to that of total tracer gas. It is seen that f_B varies between -0.05 and 0.05 with a frequency of 10 Hz, indicating an unstable gas bypassing from the draft tube to annulus. This suggests that the gas bypassing into the annulus was finally brought out of the annulus by the circulating particles. The average f_B is approximately equal to 0, much smaller than that in a conventional ICFB (usually within 15%) [13], because massive particle circulation in the mixer makes it difficult to bypass from the draft tube to annulus.

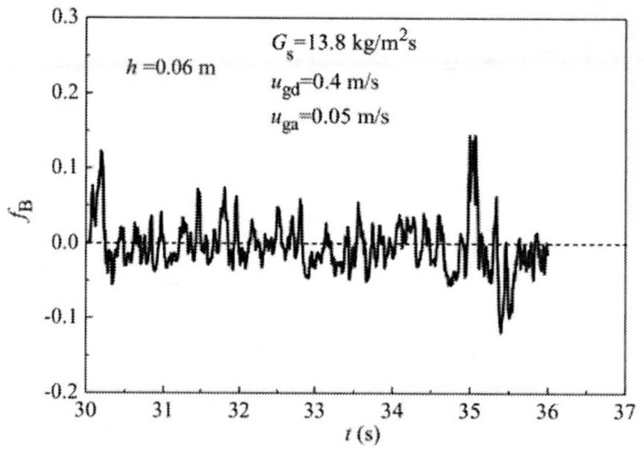

Fig. 4 Gas bypassing from draft tube to annulus in the bottom zone

As discussed previously, in the mixer there exists a pressure gradient from the draft tube to the vicinity of the slots, and then the annulus. Such a pressure difference leads to horizontal flows of particles and gas from the draft tube to the slots and then to the annulus. In the mixer the upside and downside of the slots is located at h=0.94 m and 0.86 m, respectively. Mass flux of the gas bypassing through the slots was obtained by monitoring the gas flux through the lower cross section (h=0.7 m) and the upper cross section (h=1 m) in the draft tube. Fig. 5 shows the gas mass flux under different operating conditions. The total gas mass flux in the draft tube increases with increasing gas velocity in the draft tube. With increase of the gas velocity from 0.3 to 0.45 m/s, the mass flux of the gas bypassing through the slots increases from 0.157 to 0.221 kg/m^{2}s. But the mass flux of bypassing gas does not increase linearly with the gas velocity. It is also seen that the mass ratio of the gas through the slots to that in the draft tube drops from 29.7% to 25.4%.

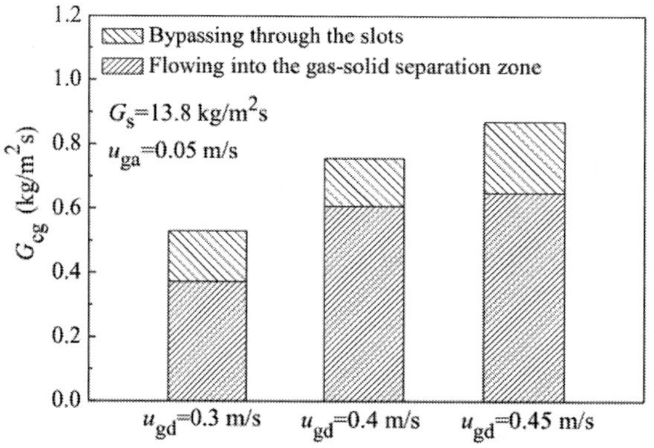

Fig. 5 Gas mass flux through the slots and the gas-solid separation zone as a function of superficial gas velocity in the draft tube

4.4 Particle mass flux

Fig. 6 shows the particle mass flux in the mixer. It is seen that the mass flux increases with increase of the superficial gas velocity in the draft tube. The mass flux through the slots accounts for about 50% of the total mass flux when u_{gd}=0.3 m/s. Because the axial particle velocity rose with increase of the superficial gas velocity from 0.3 m/s to 0.45 m/s in the draft tube, making it difficult to change the direction of the particles. Meanwhile, the radial particle velocity above the draft tube increases. Consequently, with the increase of the superficial gas velocity in the draft tube , the mass flux through the slots considerably decreased, from about 55.3% of the total mass flux (about 2.69 kg/s) at u_{gd}=0.3 m/s to only 4.8% (about 0.52 kg/s) at u_{gd}=0.45 m/s. Meanwhile, the total particle mass flux in the mixer increased from 4.87 to 10.86 kg/s as u_{gd} increased from 0.3 to 0.45 m/s.

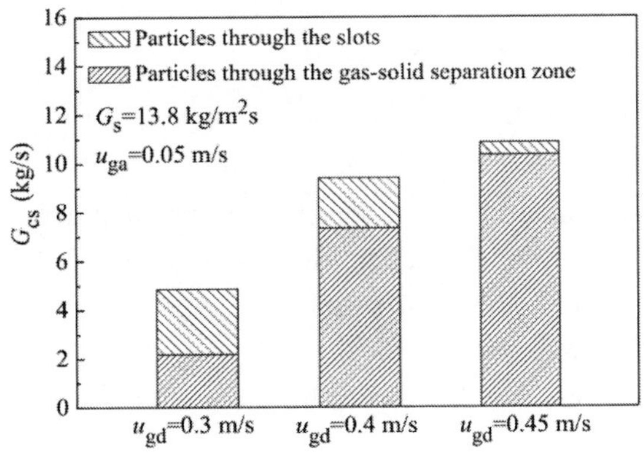

Fig. 6 Particle mass flux through the slots and the gas-solid separation zone as a function of superficial gas velocity in the draft tube

5. Conclusions

The characteristics of particle flow and mixing have a considerable influence on the mass and heat transfer, as well as the product quality. In order to further enhance the particle mixing in the ICFB, especially along the radial direction, a new-type particle mixer with slots on the draft tube was proposed in the present work. The hydrodynamics of the mixer was investigated by using a multi-scale computational fluid dynamics with the EMMS drag. The prediction in terms of cross-sectionally averaged solid holdup and solid velocity were compared to experimental data for validation. The results show that in draft tube a low pressure region can be observed near the slots and in annulus the pressure in the vicinity of the slots seems larger than that in the region between the adjacent slots, which induces horizontal particle immigration and mixing. Nearly 25.4% to 29.7% of gas in the draft tube bypasses and flows through the slots into the annulus. The ratio of the mass flux of the tracer gas bypassing into the annulus to that of total tracer gas is approximately equal to 0. With increasing the superficial gas velocity in the draft tube from 0.3 m/s to 0.45 m/s, the total solid mass flux increases from 4.87 kg/s to 10.86 kg/s, while the mass flux through slots decreases from 2.69 kg/s to 0.52 kg/s.

Notation

G_s	solid mass flux (kg/m^2 s)	G_{cg}	gas mass flux in the mixer (kg/m^2 s)
G_{cs}	solid mass flux in the mixer (kg/s)	h	height (m)

p	pressure (Pa)	R	radius of the reactor (mm)
r	radial position of measurement points (mm)	t	time (s)
u_{ga}	superficial gas velocity in the annulus (m/s)	u_{gd}	superficial gas velocity in the draft tube (m/s)
$\bar{u}$	cross-sectionally averaged axial particle velocity (m/s)		

Greek

ε_g	voidage	ε_s	solid holdup
$\bar{\varepsilon}_s$	cross-sectionally averaged solid holdup	ρ	density (kg/m^3)

References

[1] F. Niklasson, H. Thunman, F. Johnsson, B. Leckner, Estimation of solids mixing in a fluidized-bed combustor, Ind Eng Chem Res, 41 (2002) 4663-4673.

[2] L.H. Shen, M.Y. Zhang, Effect of particle size on solids mixing in bubbling fluidized beds, Powder Technol, 97 (1998) 170-177.

[3] L.H. Shen, M.Y. Zhang, Y.Q. Xu, Solids Mixing in Fluidized-Beds, Powder Technol, 84 (1995) 207-212.

[4] N. Mostoufi, J. Chaouki, Local solid mixing in gas-solid fluidized beds, Powder Technol, 114 (2001) 23-31.

[5] D.Y. Liu, X.P. Chen, Lateral solids dispersion coefficient in large-scale fluidized beds, Combust Flame, 157 (2010) 2116-2124.

[6] F. Berruti, D.S. Scott, E. Rhodes, Measuring and Modeling Lateral Solid Mixing in a 3-Dimensional Batch Gas Solid Fluidized-Bed Reactor, Can J Chem Eng, 64 (1986) 48-56.

[7] M.M. Fang, K. Luo, S.L. Yang, K. Zhang, J.R. Fan, Computational Fluid Dynamics-Discrete Element Method Investigation of Solid Mixing Characteristics in an Internally Circulating Fluidized Bed, Ind Eng Chem Res, 52 (2013) 7556-7568.

[8] J. Li, M. Kwauk, Particle-fluid Two-phase Flow: the Energy-minimization Multi-scale Method, Metallurgical Industry Press, Beijing, 1994.

[9] Z.S. Shi, W. Wang, J.H. Li, A bubble-based EMMS model for gas-solid bubbling fluidization, Chem Eng Sci, 66 (2011) 5541-5555.

[10] K. Hong, Z.S. Shi, W. Wang, J.H. Li, A structure-dependent multi-fluid model (SFM) for heterogeneous gas-solid flow, Chem Eng Sci, 99 (2013) 191-202.

[11] D. Gidaspow, Multiphase flow and fluidization: Continuum and Kinetic Theory Descriptions, Academic Press, Boston, 1994.

[12] P.C. Johnson, R. Jackson, Frictional–collisional constitutive relations for granular materials, with application to plane shearing, J. Fluid Mech. 176 (1987) 67–93.

[13] B.H. Song, Y.T. Kim, S.D. Kim, Circulation of solids and gas bypassing in an internally circulating fluidized bed with a draft tube, Chem Eng J, 68 (1997) 115-122.

514498 Power Spectral Density Analysis of Pressure Signal in 18 m Circulation Fluidized Bed Riser

Abstract:

Studies on the gas-solids flow characteristics using Power Spectral Density based on the pressure data from different axial locations in a riser of diameter 80 mm and height 18 m. Results show that the pressure signal fluctuation is mainly affected by gas-solids interaction, which included gas-solids friction, particle-particle collision and particle-particle aggregation. With the increasing solids circulating rate, frequency of particle interactions including particle-particle collision and particle-particle arrogation become strong and the pressure fluctuation also increased. It can be seen that the pressure signal fluctuation intensity decreased gradually with the increasing axial position of the riser. The low frequency but high power phenomenon nearly disappears in the fully developed zone near the top of the riser. Based on the results, it can be concluded that the gas-solids interaction is stronger at the bottom of riser. On the other hand, the gas-solid interaction is more homogeneous, and the pressure fluctuation is relatively weak in the fully developed zone.

514514 Characteristics of Pressure Fluctuations in Particle-Transport Inclined Pipe of a Circulating Fluidized Bed

Abstract:

In a large circulating fluidized bed device, the characteristics of pressure fluctuation in a particle-transport inclined pipe under different solid circulating flux were investigated based on measuring the dynamic pressure of gas-solid two-phase flow by using FCC equilibrium catalyst as the particles. The experimental results showed that with increase of particle mass flux, the flow patterns of FCC particles above and below the butterfly valve in the inclined pipe both successively changed, and the fluctuation amplitude of dynamic pressure increased. The standard deviation of the dynamic pressure below the butterfly valve possessed a linear relation with the particle mass flux, which could be used to characterize the real-time dynamic pressure in the inclined pipe. By wavelet analysis, it is indicated that there exists a main frequency caused by inclined pipe flow structures in the pressure fluctuations above and below the butterfly valve, besides, there is a secondary frequency in the pressure fluctuation below the butterfly valve caused by gas turbulence and gas-solid interactions. With the increase of particle mass flux, the wavelet energy ratio of the secondary frequency gradually decreased. The fluctuation pressure caused by discharge in the inclined pipe is a low frequency pressure and the excitation source of inclined pipe vibration.

508323 Dynamic Modeling of Attrition and Reactions in Circulating Fluidized Bed Reactors

Abstract:

Fluidized bed reactors are widely applied in chemical processes due to the excellent gas-solid contact. A problem that occurs in fluidized beds is the mechanical stress that particles are exposed to during the fluidization, which causes particle attrition. Additionally, fluidized beds are often operated in batch mode, which will cause ever-changing operating conditions with time. This makes fluidized bed modeling especially challenging, because the distributed particle properties, like size, wetness or reaction state, have to be tracked over the whole course of the simulation.

At the Institute of Solids Process Engineering at Hamburg University of Technology, novel process simulation software named DYSSOL is developed, particularly tailored for solids processes. The software is based on the flowsheeting concept. This means that chemical unit operations, like for example chemical reactors or separators, are put into process modules, which are interconnected into a process flowsheet and then mathematically solved. To make dynamic simulations available, new dynamic models had to be proposed for the involved equipment. The models themselves are based on empirical and semi-empirical correlations for hydrodynamics and kinetic particle reaction models are integrated.

The dynamic flowsheeting software was then used to simulate a system of interconnected fluidized bed reactors, which are used for Chemical Looping Combustion of solid fuels. This process is especially interesting, because heterogeneous reactions take place. Further, the process is a reactor-regenerator system causing the reactive solid to be oxidized and reduced and, hence changing its conversion state. Attrition plays a major role for the cost effective operation of the process.

In a small lab scale bubbling bed reactor, the jet and bubbling attrition rate constants for the solid oxygen carrier were experimentally determined under ambient conditions. In a small stand-alone cyclone the cyclone attrition was investigated. Furthermore, the reaction kinetics of the solid material with fuel gases was experimentally determined. All this information was included into the flowsheet simulation of a pilot scale circulating fluidized bed reactor, which is actually operated in our institute to deliver validation data.

The dynamic flowsheet simulation of the pilot plant, which was enhanced with kinetic and attrition information of smaller scale facilities, was conducted over 1000 seconds. A fuel load change was simulated and the chemical conversion was tracked. Over the whole simulation, the attrition and small particles leaving the reactor system was also covered. When comparing the simulation with experiments it could be shown that the gas conversion was sufficiently calculated and the conversion state of the oxygen carrier tracked correctly. The attrition modeling, however, with its parameters measured had to be adapted, so that it fitted the outlet particle flows measured under reaction conditions.

The financial support of DFG (Deutsche Forschungsgemeinschaft) within the priority program SPP 1679: â€œDynamic simulation of interconnected solids processesâ€ is gratefully acknowledged; grant number HA 6935/2-1.

514599 Analysis of FCC Cyclone Fault Diagnosis Technology Based on Particles Information

Abstract:

Cyclones are key equipment in fluid catalytic cracking unit (FCCU), which are used for separating the catalyst particles from oil gas or flue gas. Hence, it is very necessary to carry out condition monitoring and fault diagnosis for the running cyclone. The properties of particles vary at different positions in the particle circulation loop, such as size distribution, morphology and element composition, etc. In consequence, the particles contain much information about operation. Once the cyclone operates abnormally, the information of run-off catalyst particles appear abnormal too, and there exists a corresponding relationship between them. So, taking the catalyst particles as the carrier of fault information, this paper put forward cyclone fault diagnosis technology by extracting catalyst particle information, such as particle packing density, particle size distribution, particle micrographs etc, to build the relationship model between the cyclone fault and catalyst particle information.

514600 The Multi-Hole Throttling Effect and the Erosion Characteristics of the High Pressure Natural Gas

Abstract:

Based on Euler-Lagrange multi-phase flow numerical simulation method, the single hole and multi-hole throttling process of the high pressure natural gas was studied by using the RNG k-$\hat{I}\mu$ turbulent model, the Sutherland Viscosity Law compressible fluid viscosity model, and the advanced erosion model numerical simulation. The calculated mass flow variation law in the throttle choke is comparable with the theoretical calculations, which verified the adopted models and calculating methods. And then the throttling effect and the erosion characteristic of the high pressure natural gas was studied to check its specific law with different throttle holes by using the advanced erosion model. The results showed that, high pressure natural gas can reduce pressure and expansion after flowing through the single hole and multi-hole throttle choke, in which lead a high-speed jet region came up near its outlet. But the maximum jet velocity of multi-hole throttle choke was significantly lower than that of the single hole. At the same time the multi-hole uniform jet may reduce the vortex entrainment and the wall erosion damage, in which can be easier to mix the cold core and achieve stable pressure recovery at the role of the multi-jet mixing effect. Therefore, the multi-hole throttling can effectively restrain the erosion damage and the gas hydrate formation probability.

514601 Numerical Simulation of Flow Field In a Gas Pipe Distributor of the FCCU Regenerator

Abstract:

The flow field in the pipe distributor used in regenerator of FCCU (Fluid Catalytic Cracking Unit), was analyzed based on the numerical simulation. Firstly, the distributing gas performance of the pipe distributor was investigated. The result shows that the pressure increases along the branch pipe of the pipe distributor and the axial flow rate decreases in the branch pipe, which leads to the nozzle flow rate increasing gradually. Furthermore, an obvious bias flow was found in the inlet of branch pipe, which brings some nozzle flow rate less than the designed flow rate and causes an uneven gas distribution in the fluidized bed as well. Meanwhile, the momentum equation with variable mass flow was applied to analyze the flow in the pipe distributor. The flow type in the branch pipe belongs to the "momentum exchange controlling model". The pressure in the branch pipe always increases from the inlet to the end, and the pressure distribution is not consequentially uniform. The uneven gas distribution of the pipe distributor results in fluidized gas non-uniformity in the bed and erosion in the branch pipe and nozzle. Finally, according to the pressure characteristics in fluidized bed system, relationship between the pressure drop of branch pipe and pressure drop of nozzles, as well as of the critical nozzle pressure drop and the designed nozzle pressure drop, were discussed. The gas distributor performance could be optimized by means of the structural improvement of branch pipe.

514602 Full-Loop Simulation of Gas-solids Flow in Circulating Fluidized Bed of FCC System

Abstract:

Circulating fluidized beds are widely used in petrochemical, coal, metallurgy, and other industrial fields. A typical circulating fluidized bed (CFB) includes the riser, gas-solid separator, downer and solids flow control device. To obtain detailed information and get better understand of gas-solids flow in CFB, computational fluid dynamics (CFD) had been applied extensively in many systems such as circulating fluidized bed combustion (CFBC) systems and fluid catalytic cracking (FCC) systems. In most studies, the riser, the core part in CFB system, was isolated as a single simulation object by applying proper boundary conditions. However, there are some limitations of riser-only simulation, without consideration of dynamics in solids circulation rate and effects of other part of CFB system. Therefore, the full-loop simulation is a more promising way to describe complex gas-solids interaction in CFB by treating the whole CFB system as a simulation object. Nowadays, most of current full-loop simulations were focused on CFBC systems, there was little work about FCC system, with different type of particles and recycle mechanism from CFBC system.

In this work, both full-loop and riser-only simulations were carried out by TFM approach to investigate a pilot-scale gas-solid circulating fluidized bed of FCC system which has been set up in our group. The simulation condition was chosen according to experiments with different superficial gas velocity, valve opening and fluidizing air flow rate. Detailed comparison between full-loop and riser-only numerical simulations has been conducted with respect to the hydrodynamics inside the riser, which showed full-loop simulation had advantages on predicting dynamics of solids circulation rate and the gas-solids flow behavior. Distribution of pressure, solids holdup and velocity in system was investigated which was helpful to understand the hydrodynamics in CFB. The effects of operating condition were then studied, which revealed the system stability mechanism.

506758 Immobilization of Sulfur-Oxidizing Bacterium, *Thioalkalivibrio Sp.* D301 on Magnetic Nanoparticles and Biodesulfurization

Abstract:

There is an urgent need to develop desulfurization methods for halo-alkaline wastewater. *Thialkalivibrio sp.* D301, an obligatory chemolithoautotrophic, haloalkaliphilic, sulfur-oxidizing bacterium, was used in the removal of sulfide in wastewater. We developed a process to immobilize D301 cells by using superparamagnetic Fe_3O_4 nanoparticles (NPs) under halo-alkaline conditions. Fe_3O_4 NPs were synthesized by coprecipitation of ferric and ferrous salts in the presence of N_2 gas. TEM image shows the spherical shape of Fe_3O_4 NPs with an average diameter of 10 to 15 nm. The particles had superparamagnetic properties. The saturation magnetization (Î´$_s$) of the Fe_3O_4 NPs was 73.985 emu g$^{\text{â}ˆ\text{'}1}$ with zero remanence and coercivity. Analysis of zeta potentials and hydraulic diameters showed that the Fe_3O_4 NPs and bacterial cells interact mainly through the nano-size effect. The NPs-coated cells have the same sulfur oxidization activity as that of free cells. Experiment results showed that the immobilized cells still retained >90% of their sulfur oxidization activity after six times of reuse. The cells coated with Fe_3O_4 NPs did not interact through electrostatic attraction. The mechanism of NPs adsorption mainly involved the nano-size effect. Magnetic immobilization is a convenient technique for HA-SOB immobilization under halo-alkaline conditions and is a promising technology for large-scale industrial applications.

507944 Simulation of Bulk Solids and Granular Systems By Using Combined Discrete Element Models

Abstract:

Granular materials can be found in many industrial processes from chemical to energy, agriculture, mining, pharmacy and many other fields. They are almost always opaque, making the acquisition of experimental data below the surface difficult. DEM (Discrete Element Method) simulations are an effective way to study the physics of granular materials because they can simulate the micro-dynamic behavior of granular material on an individual particle scale and the motion of every particle can be traced. Since DEM was first considered, many modified discrete element models have been developed. However, most existing DEM simulation are just for spherical particles or a specific type of non-spherical particles. In this work, a combination of several types of discrete element models including spherical models, super-ellipsoids, multi-sphere models (with no interaction and deformation between sub-spheres), and bonded-spheres models (with interaction and deformation between sub-spheres) were used to simulate the flow of granular materials with many different shapes. Several simulation cases including the flow of particles in tumblers, blenders, and other devices were performed to validate the modeling approaches. The simulation results show that it is practical and efficient to combine different discrete element models together for simulating both spherical and non-spherical particles in bulk solids and granular systems.

515703 Material Optimization of Perovskite Films using High Throughput Synthesis and Multi-Dimensional Analysis

Abstract:

Over the past years, Perovskites received special attention in the field of renewable energy as they are one of the prime candidates for energy storage and conversion materials. Perovskites act as the active light harvesting layer of a solar cell and provide excellent performance.[1] A common synthesis strategy is through the liquid phase under inert atmosphere.[2,3] However, this comes with the challenge of a huge parameter space that in principle would have to be first synthesized, then characterized and finally optimized based on the therefrom deduced knowledge.

To overcome these limitations, we follow a fully automated approach, which allows automated high throughput (HT) synthesis and HT characterization of perovskites with high accuracy. As an efficient route to explore the parameter space, we use design of experiments (DoE) to find optimized settings for process parameters such as synthesis and annealing time and temperature, but also material aspects like the molar ratio of the reactants.

For all experiments, a Chemspeed© technologies (CST) swing XL platform with a liquid handling tool and a heatable shaking rack was used. The robotic platform is enclosed within an Mbraun© glovebox with inert atmosphere (Nitrogen). After precursor preparation, deposition was realized by a liquid handling tool on glass substrates inserted in a customized aluminum 24 well plate, which is then transferred onto a hot plate for film formation.

In a first step, after annealing at different temperatures, a characterization cascade was developed and applied that contains the subsequent elimination of samples that exhibit undesired properties and side products. Fast absorption analysis, photoluminescence (PL) measurements and finally X-ray diffraction (XRD) were performed to identify optimum chemistry and process parameters. Although this is a very reliable procedure, it is still comparatively demanding in terms of the time effort. Therefore a more efficient, purely optical approach by which the experimental parameter space is only explored in terms of rapid emission analysis was developed in a second step. Noteworthy, both strategies led to the same optimum conditions, however, at significantly reduced time effort.

In conclusion, our work shows how by the combination of automation, DoE and efficient characterization, optimum conditions for new materials like perovskites are rapidly identified. This opens the door towards faster material development which is important whenever new perovskite structures, e.g. lead free materials for industrial applications, are strived for.

Literature

1. Green, M.A.; Ho-Bailie, A.; Snaith, H.J., "The emergence of perovskite solar cells", Nature Photonics 2014, 8, (7), 506-514.
2. Baikie, T.; Fang, Y.; Kadro, J.M.; Schreyer, M.; Wei, F.; Mhaisalkar, S.G.; Graetzel, M.; White, T.J., "Synthesis and crystal chemistry of the hybrid perovskite (CH 3NH3)PbI3 for solid-state sensitised solar cell applications", J. Mater. Chem. A 2013,1 , 5628-5641.
3. Yan, K.; Long, M.; Zhang, T.; Wei, Z.; Chen, H.; Yang, S.; Xu, J., "Hybrid Halide Perovskite Solar Cell Precursors: Colloidal Chemistry and Coordination Engineering behind Device Processing for High Efficiency", J. Am. Chem. Soc. 2015, 137, (13), 4460–4468.

509509 Solid Liquid Separation Via Particle Flow Instability

Abstract:

A new separation strategy using novel particle-flow-instability physics is successfully developed for a *difficult-to-separate suspension* in which there is some combination of a small density difference between solid and liquid, high viscosity, and small-sized particles. The method we propose here requires no dilution for medium, high or extremely high viscosity slurries and can produce effective solidâ€"liquid separation with low capital and operational inputs. The results have relevance for many biotechnological and/or chemical processes.

509543 Milling and Grindability Assessment of Pharmaceutical Materials

Abstract:

Milling is one of the most important unit operations in the pharmaceutical industries. Most of the Active Pharmaceutical Ingredients (APIs) and excipients are produced by crystallisation, and grown to a sufficiently large particle size suitable for isolation and drying. Therefore most of the materials have to be milled to desirable particle sizes. Milling is influenced by a combined effect of milling dynamics (input power, mill geometry, etc) and material properties. Considerable research has been carried out on analysing the particle behaviour in various mills. Nevertheless, the prediction of the milling rate and milled product properties (e.g. product size distribution) is still very challenging.

In this work, the milling behaviour of several pharmaceutical materials in the PicoPlex mill (pin mill unit of the Picoline, Hosokawa Alpine) has been studied. A combination of experiments and Distinct Element Method (DEM) simulations is used to explore the dynamic behaviours of the materials in the mill. The results are then used to validate a novel approach of grindability assessment of pharmaceutical materials using the Scirocco disperser of the Malvern Mastersizer 2000 (Bonakdar et al. 2016). Based on this method the ease of grindability of the materials in the mill can be predicted.

Bonakdar, T., Ali, M., Dogbe, S., Ghadiri, M., and Tinke, A. (2016). A method for grindability testing using the scirocco disperser. *International Journal of Pharmaceutics,* 501: 65-74

507857 Bi-Directional Thermal Control of Twin Screw Granulation Process Via a Specialised Annular Heat Pipe

Abstract:

The continuous process of twin screw granulation forms a hotbed in pharmaceutical research as it is expected to succeed the current batch production of granules. In essence, the scale up of the material output is a function of production time, instead of equipment size (for batch granulators). This way, a â€œjust-in-time productionâ€ of the desired amount of material can be performed according to specific needs. In this work, a twin screw granulator was utilised in the granulation of both the placebo and active pharmaceutical formulations. Subsequently, the same equipment was used in granulating ceramic raw material powders for tile pressing. During processing, the spatial constraint in the barrel and screw compartments caused friction, resulting in excess heat generation. For the thermally-sensitive pharmaceutical formulations, this must be controlled to avoid denaturation of the granule products. A common solution is by using flowing cool water, but a good control has been proven to be difficult to achieve as different granulation zones generate different heat levels (kneading blocks > conveyer blocks). Heat pipe is a passive and highly effective thermal conductor and stabiliser. A carefully-designed annular heat pipe, filled with a small amount of methanol was successfully employed to stabilise the temperature in the barrel space, cooling the granulating material before it reached its critical temperature. For the ceramic powders, the methanol filling of the heat pipe was replaced by water and it was then heated up to provide an in-situ drying operation as the powders granulating. The in-situ drying in this instance, reduces energy consumption for the downstream drying processing as the product granules contain less moisture to be removed. The annular heat pipe, with its bi-directional operation of heating and cooling is a technically viable temperature controller for twin screw granulators. Compared to the established active cooling methods, the heat pipe provides a passive alternative, with only a discreet amount of fluid required to last its operational lifetime, thereby providing an economic thermal control solution.

Mixing Grains with Different Elongation in a Rotating Drums

Claudia Piacenza[1,2], Marco Marconati[1], Colin Hare[1], Andrea C. Santomaso[2], Marco Ramaioli[1,]*

[1]*Department of Chemical and Process Engineering, Faculty of Engineering and Physical Sciences, University of Surrey, Guildford GU2 7XH, UK*
[2] *APT Lab, Department of Industrial Engineering, University of Padova, Padova, Italy*

[*] *Email: m.ramaioli@surrey.ac.uk*

Abstract

Mixing grains with different properties is a remarkably challenging process, relevant to many industrial applications. Rotating drums have been used extensively as model systems to study granular media flow and mixing and segregation. Numerous studies considered the mixing of grains with different sizes, but only few studies considered shape and elongation, which have been already identified as important characteristics affecting packing and inducing segregation. In this contribution, the mixing of binary mixtures of grains having the same volume, but different elongations was studied experimentally. The mixing dynamics of a layered granular medium was characterized in a rotating drum, highlighting the impact of the grain shape, of the drum angular speed and of the drum filling ratio. A mixed or segregated state is reached very rapidly, but experiments are continued to verify that the state reached is a steady state. The experiments demonstrated that below a critical difference of elongation, the grains can be mixed effectively. Conversely, when grain elongations are very different, a central core is formed, rich in the more elongated grains. In this case only limited mixing can be achieved. These results can guide the formulation of mixtures of grains with different elongations.

1. Introduction

Mixing is a process operation aimed at obtaining a homogeneous system starting from two or more components. On the contrary, when there is a lack of homogeneity within the mixture segregation occurs. Segregation phenomena represent a relevant problem in the case of mixing solid particles having different physical properties such as particle size, density, shape or surface roughness. Differently from liquids and gases, solid particles are not thermal activated system [1] and so in order to obtain a certain degree of mixing they need to be mechanically activated. Typically three mixing mechanisms are largely recognized:

- dispersion, analogous to diffusion in fluids, it involves single particles that randomly move due to collision between each other;
- convection, it involves a portion of the bulk material, with a given direction and velocity;
- shear mixing: it is peculiar to granular solids and it is a combination of the two previous mechanisms.

Powder mixers are relatively simple machines of low capital cost. They are classified into two types: tumbling mixers and agitated mixers. Tumbling mixers are the simplest. A totally enclosed vessel is rotated about an axis causing the particles within the mixer to tumble over each other on the mixture surface. Tumbling vessels provide low-shear environments. Convective mixers instead are characterized by the action of impellers or paddles to move the powders around and generate a well-mixed product. Convective mixers provide more shear into the mixture and can be used with cohesive materials.

Rotating drums belong to the tumbling mixers. They are extensively used in the chemical and process industries. With respect to mixing processes, rotating drums are often used to study and understand mixing and segregation phenomena; as a result, many experimental and numerical studies were developed on granular behavior in rotating drums [1-3]. Different granular flow regimes can occur in the drum, each one with its specific flow behavior. In particular, different flow regimes can be assessed by using the dimensionless Froude number, $Fr = \omega^2 R/g$ [4], where ω [s^{-1}] is the drum rotation speed, R [m] the drum radius and g [m/s^2] the gravitational acceleration. With $10^{-4}<Fr<10^{-2}$ and filling level $\varphi>0.1$ rolling regime takes place. This motion is characterized by a uniform, steady flow of a particle layer on the surface (cascading or active layer) while the larger part of the bed (plug flow region) is moved upwards by solid rotation with the rotational speed of the wall. A uniform and good mixing is achievable with this type of motion, but when particle size distribution is very broad segregation can appear. The rolling bed is preferred, since is the simplest regime providing favorable conditions for heat transfer and high quality of the product even when mass flow rates are large.

When particles to be mixed have different properties such as different size, different density or different shape segregation typically occurs and particles tend to separate. [5, 6]. Difference in size is by far the most important cause of particle segregation. Particle size distribution influences also strongly bulk properties such as packing fraction and bulk modulus [7].

Different size segregation mechanisms can be identified:

- trajectory mechanism: according to the Stokes' law the velocity of fine particles is lower than that of big particles. This can cause segregation when small particles move through air;
- percolation mechanism: if a bulk material is disturbed, a rearrangement in the particle packing occurs and for fine particles it is easier to fall down in the inter-particle voids;
- inertial mechanism: in pouring and filling operations bigger particles tend to flow selectively to the bottom of the pile since they have a higher momentum;
- elutriation mechanism: in presence of air the very fine particles can remain in suspension while larger ones tends to settled. This can occur in charging storage vessels or hoppers.

A homogeneous binary granular mixture, with differences in size and density, in a two-dimensional rotating drum typically gives radial segregation. Considering a rotating drum operating in the rolling regime filled with particles of different sizes, bigger particles mainly tend to the edge of the system, while smaller particles accumulate to the centre of the drum. The phenomenon is fast and the rotational speed of the drum seems not to affect the phenomenon; the size ratio of the particles is, instead, the parameter that mainly affects size segregation phenomena. Most of the experimental investigations about segregation in the rotating drum involved spherical or nearly-spherical particles; the experimental studies of the flow of non-spherical particles are, in fact, rather limited [6,8]. The shape of the particles, however, strongly influences the motion in the system. In particular, the static and dynamic angles of repose of non-spherical particles are found to be very different from the ones measured for spherical particles suggesting that differences in particle mobility can affect the segregation process [9].

2. Materials and methods

2.1 Particles

Both spherical and non-spherical particles were used for the experiments. Their size (diameter d for spheres, major and minor axis for ellipsoids a and b respectively) , aspect ratio AR, projected area A_p, volume V, and true density ρ are reported in Table 1.

Table 1. Physical properties of the particles

Type	d [mm]	a [mm]	b [mm]	AR [-]	A_p [mm^2]	V [mm^3]	ρ [g/cm^3]
Spheres S	8	-	-	1	50.3	268	1.09
Ellipsoids E1	-	10.48	6.98	1.5	57.5	267	0.96
Ellipsoids E2	-	18.44	5.26	3.5	76.2	267	0.98

Spherical particles with size 8 mm were used for the experiments. The ellipsoids were 3D printed with *AR* equal to 1.5 and 3.5. They were printed with the Ultimaker 3D printer and made of polylactic acid (PLA). The software used to realize the 3D design of the particles used in this work is FreeCAD, which is an open-source parametric 3D modeller allowing to obtain STL files.

The true density of all the particles used for the experiments was measured by gas pycnometry (Multivolume Pycnometer 1305, Micromeritics). The measured density of the spheres was very close to the typical density of the nylon 66 (1.14 g/cm^3). The PLA ellipsoids had a density that was lower than the typical density of the PLA (1.24 g/cm^3). This is because the particles were printed with a set infill density of the 20%, so they where partially empty inside.

2.2 Tumbling mixer

The quasi-2D drum was made of two transparent polycarbonate disks with a diameter of 30 cm and a thickness of 6 mm; between them a wooden spacer was placed, with the same external diameter, an internal diameter of 25 cm and a thickness of about 9.1 mm. The spacer allowed to create an internal volume of finite thickness separating the two polycarbonate disks.

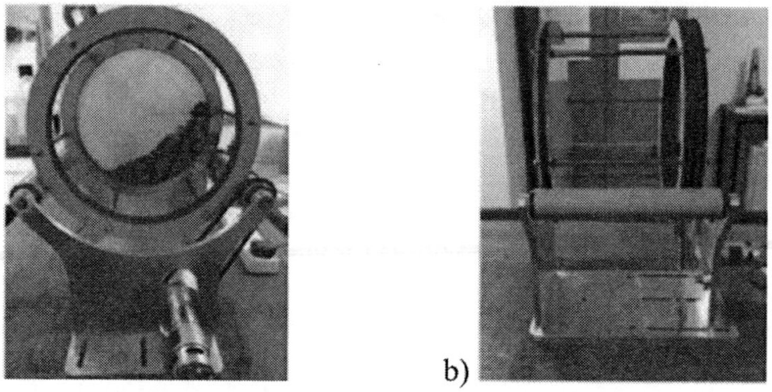

a) b)

Figure 1. a) Front view and b) side view of the rotating drum (the drum is on the left side of the rigid structure).

The internal lateral wall of the volume was covered with sandpaper in order to provide friction and avoid undesired slip of the particles. To keep the drum in vertical position and allow its rotation, three additional wooden spacers were used, at a distance of about 16.3 cm. This rigid structure could rotate at variable rotational speed thanks to a motor connected to a voltage supplier. Two rollers mounted on a rigid support made the rotation of the drum possible; a pulley transmitted the motion to the driving roller, while the second roller was passive. The rotational speed could be changed by varying the motor voltage. The structure could be removed from the rollers and by unscrewing the bolts, the drum could be easily opened so that to load/unload the particles.

2.3 Image acquisition

Frontal pictures of the drum were taken by a color camera and analyzed by the image processing software ImageJ [10]. To avoid the reflection of the camera and other objects on the front transparent disk, a squared LED panel was used which allowed to have a dominant diffused light illuminating the drum. The LED had a power of 8 W and a color temperature of 4000 K; it was placed behind the drum and opposite to the camera. A circular sheet of paper, which was attached to the drum between it and the LED, diffused the light properly. The camera was a Basler ace camera – model acA2000-165uc, with a maximum resolution of 2040 × 1086 pixels. The focus and the diaphragm of the camera could be adjusted manually.

2.4 Experimental plan

The attention was focused on mixtures of particles with different shapes, since segregation phenomena due to size had been widely investigated in the past. Three mixtures were considered, with two possible initial top-bottom segregated configurations. Filling level and rotational speed where changed also as detailed in Table 2.

Table 2. Experimental tests performed with differently shaped particles.

AR 1 [-]	*AR* 2 [-]	ω [rpm]	φ [-]	Init. Config.
1	1.5	4.22	25%	E1/S
1	1.5	4.22	25%	S/E1
1	1.5	4.22	47%	E1/S
1	1.5	4.22	47%	S/E1
1	1.5	11.51	25%	E1/S
1	1.5	11.51	47%	E1/S
1	3.5	4.22	25%	E2/S
1	3.5	4.22	25%	S/E2
1	3.5	4.22	47%	E2/S
1	3.5	11.51	25%	E2/S
1	3.5	11.51	25%	S/E2
1	3.5	11.51	47%	E2/S
1.5	3.5	4.22	25%	E2/E1
1.5	3.5	4.22	25%	E1/E2
1.5	3.5	4.22	47%	E2/E1
1.5	3.5	4.22	47%	E1/E2

*AR*1 and *AR*2 indicate the aspect ratios of the particles in the binary mixture. It is reminded that the spheres (*AR*=1) and the two types of ellipsoids (*AR*=1.5 and *AR*=3.5) had the same volume. The E1/S initial configuration indicates a segregated configuration with short ellipsoids above spheres (this was assumed as basis configuration); the S/E1 configuration is inverted with spheres above short ellipsoids. A similar convention is applied to the experiments considering mixtures of ellipsoids. All the experiments were performed with binary mixtures with a composition of 50% and 50% on a number (and hence also volume) basis. The filling level φ of the 25% was obtained with a total number of particles equal to 200; with 320 total particles the filling level of the drum was equal to 47%.

2.5 Image analysis procedure

Image analysis techniques were used to identify the particles on the basis of their different shape. The procedure was automated using customized ImageJ macros. Contrast and brightness of the images were first adjusted. Color images were then set to gray scale and segmented by applying a threshold. The operation was carried out twice to separate light and dark particles with different thresholds. Watershed segmentation was then used to separate touching objects. This last operation was more difficult for the particles at the border of the drum; in some case, in fact, the ellipsoids were erroneously identified as spheres. Also, the presence of the border of the drum complicated the watershed operation.

The centre of mass of each particle ($x_{m,i}$ and $y_{m,i}$) was estimated and the radial distance (d_i) from the centre of the drum (x_c, y_c) calculated as:

$$d_i=[(x_c-x_{m,i})^2+(y_c-y_{m,i})^2]^{0.5}$$

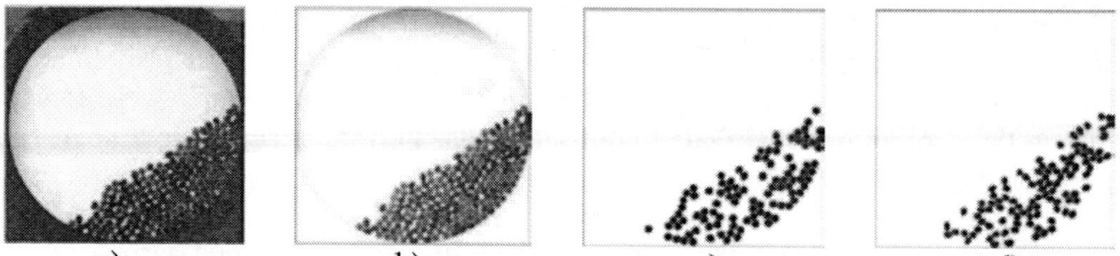

a) b) c) d)

Figure 2. Example of image processing steps. a) Original colour image; b) grayscale conversion and circle cut from the original picture; segmentation applied to c) light particles and d) dark particles.

The mean value of distance (d_m) for the spheres and for the ellipsoids was then calculated as:

$$d_m = \Sigma\, d_i / N$$

Only for very elongated ellipsoid (AR=3.5) the watershed operation failed resulting in over-segmentation errors. Single particles could not be identified unequivocally in this case; only one center of mass for all the elongated ellipsoid was estimated and the average distance (d_m) of the particles from the drum center calculated.

3. Results and discussion

Figure 3 shows two typical experiments, carried out at the lowest rotational speed and filling ratio, with ellipsoids (AR=1.5 at left and AR=3.5 at right) on top and spheres on bottom in the initial and final configuration. It is evident even by visual inspection that the final state presented some level of radial segregation and that a different final segregation intensity was reached in the two systems.

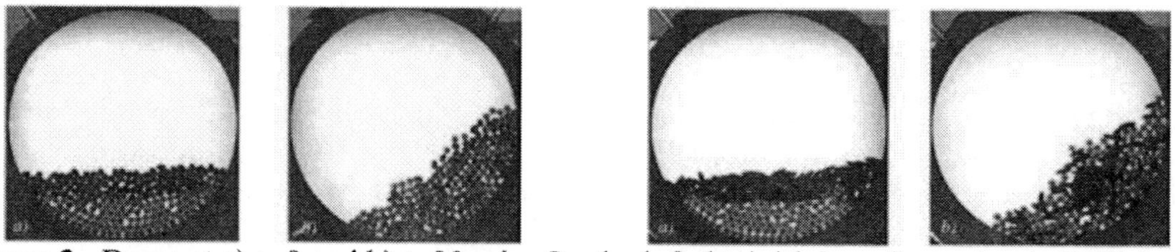

Figure 3. Drum at a) t=0 and b) t=30 min. On the left the initial configuration is E1/S (using ellipsoids with AR=1.5) and on the right the initial configuration is E2/S (AR=3.5). In both experiments ω=4.22 rpm and φ= 25%.

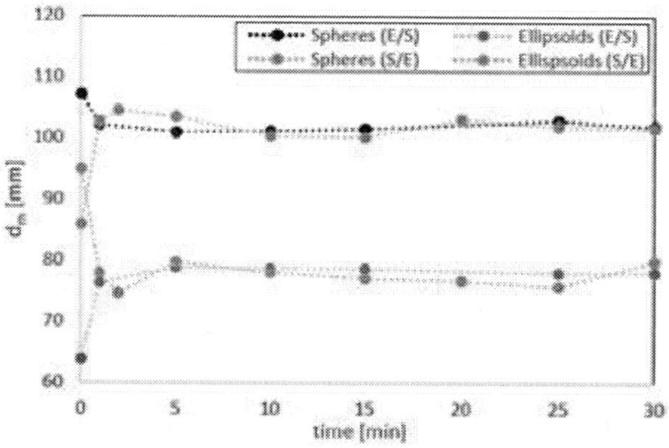

Figure 4. Evolution of the mean radial distance vs. time for the mixture of spheres and ellipsoids with AR=3.5. The initial configuration E2/S is compared to the initial configuration S/E2.

A more quantitative analysis was obtained by plotting the trend of the mean radial distance (d_m) of the two species at different time instants. As a typical example of the system evolution, the case of E2/S

is shown in Figure 4. The d_m reached a steady value after a short time. Dedicated experiments on a shorter time scale (2 minutes, not shown here) revealed that steady state value for d_m was reached after 30s for mixtures with ellipsoids E1 (AR=1.5) and after 60s for mixtures with E2 (AR=3.5).
At steady state, the radial distance was in both cases smaller for the ellipsoids, meaning that they mainly remained concentrated on top or at the center of the granular bed while the spheres mainly concentrated at the periphery of the drum (close to the drum wall). This tendency was much stronger for ellipsoids E2. So a slower segregation dynamics compensated by a larger segregation intensity was observed with the more elongated particles.
As Figure 4 shows, inverting the initial configuration from E/S to S/E did not affect the dynamics and the final result in all the experiments performed with these particles. Initial configuration resulted to have no effect on the segregation propensity of the mixtures analyzed.

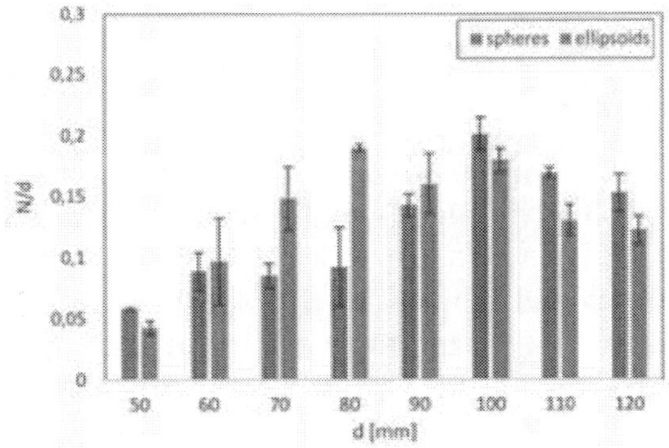

Figure 5. Distribution of the radial distance of spheres and ellipsoids with AR=1.5, obtained by averaging the results of three repetitions. Other conditions are t=30 min, ω=4.22 rpm, φ= 25%.

A more in-depth analysis can be carried out considering the distribution of the radial distance of the spheres and ellipsoids separately. The radial distance from the centre of the drum (d) was divided in classes of 10 mm and the number of each type of particles (N) in each class was computed. To obtain the distribution, N was normalized on the radial distance as N/d and plotted versus d. In the case of completely homogeneous mixtures and assuming that the medium takes the shape of a circular sector, N/d should be constant and equal for the two types of particles in each class. The peak of the distribution of the ellipsoids was found at lower distance than the peak of the distribution of the spheres. Moreover, at lower distance, the number of ellipsoids was higher than the number of spheres, especially for distances between 70 and 90 mm; at higher distance (closer to the drum wall), instead, the number of spheres was higher than the number of ellipsoids, thus confirming the previous analysis based on d_m.

Also, binary mixtures made of the two types of ellipsoids E1 and E2 (AR=1.5 and AR=3.5 respectively) were taken into account at the same filling levels of the previous experimental tests and at the lower rotational speed. Stable segregation was again observed with intermediate results with respect to the two previous cases (S+E1 and S+E2). So as a general comment on these experiments, stable segregation was always observed when mixing particles having same volume but different AR. The intensity of segregation increased with increasing the difference in AR between the particles. In particular particles with larger AR tended to accumulate at the center of the granular bed, while particle with lower AR accumulated at the periphery. Segregation occurred in the surface active layer where the more elongated particles, because of their reduced propensity to flow (particle rolling was prevented by their elongated shape), were systematically overtaken by the less elongated (and faster) particles. In this way a spontaneous separation of particles occurred in the active layer, with those having a lower AR arriving first at the drum wall, while those having an higher AR remaining trapped in the core.

3.1 Effect of filling level

The comparison of the behavior of the mixtures at the two filling levels can be carried out by normalizing the mean radial distance d_m of the two types of particle on the mean distance of all the particles d_c, calculated as $(d_{m,\ spheres}+d_{m,\ ellips.})/2$. The trends of d_m/d_c are shown in Figure 6

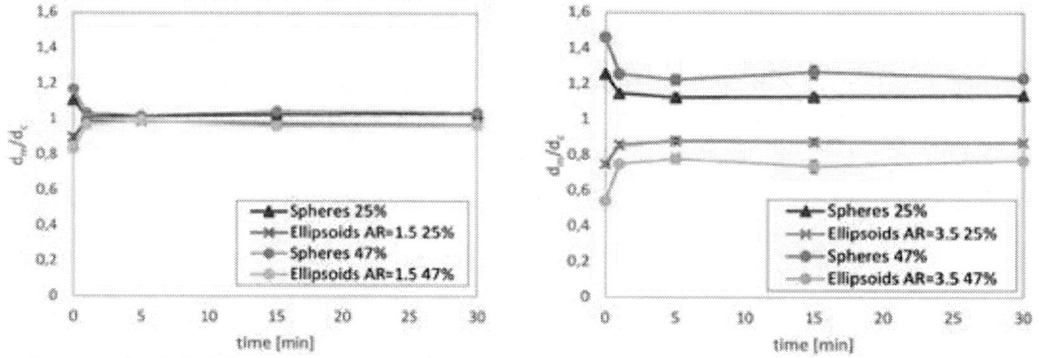

Figure 6. Trend of d_m/d_c vs. time for the mixture of spheres and ellipsoids with *AR*=1.5 (left) and AR=3.5 (right) at two different filling levels.

In the case of mixture of spheres and ellipsoids with *AR*=1.5, the curves at different filling levels overlap, showing that the behaviour of the mixture was not affected by the filling level of the drum. In the case of the other two mixtures (S+E2 and E1+E2, not shown), the curves did not overlap , especially for the case S+E2 shown in Figure 6 (right). An explanation can be found in terms of bed geometry. When increasing the filling level (below 50%) the length of the surface active layer increased as well. This gave more chances for particles to separate, in particular when the difference in *AR* was large enough (S+E2 and E1+E2). However for the system S+E1 where the difference in *AR* was the smallest, the difference in active layer length in the two cases was not sufficient to create observable differences in the segregation pattern.

3.2 Effect of rotational speed

In addition to experiments carried out at 4.2 rpm, tests were repeated at 11.5 rpm with spheres and ellipsoids (25% filling level). Figure 7 shows that rotational speed did not affect the segregation patterns, at least in the range of the analyzed speeds.

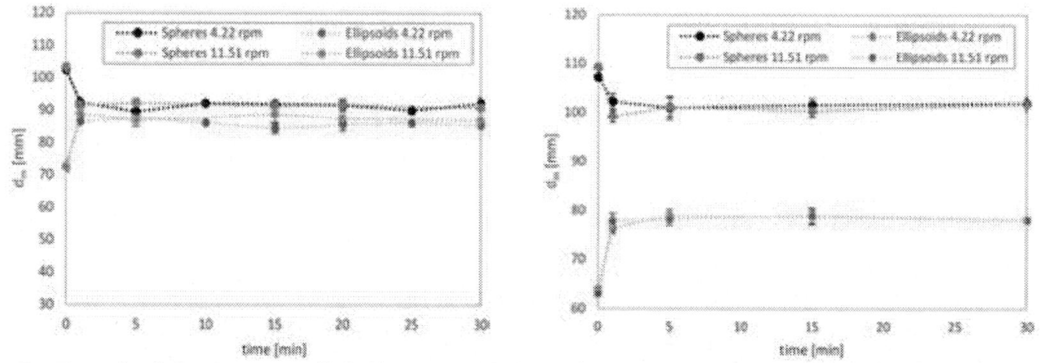

Figure 7. Trend of the mean radial distance vs. time for the mixture of spheres and ellipsoids with *AR*=1.5 (left) and *AR*=3.5 (right) at two different rotational speeds.

Conclusions

From the experimental tests described in the previous paragraphs, it was found that binary mixtures of particles with different shape but with the same volume do not mix uniformly and segregation was always observed. In particular, the particles with lower aspect ratio tended to stay closer to the drum walls, while the more elongated particles were found preferentially at the centre of the granular bed. The most significant case was represented by the mixture of spheres and ellipsoids with AR=3.5; for this mixture, in fact, the segregation was more evident and the non-spherical particles clearly accumulated in the centre of the granular bed. The same final state was obtained regardless of whether the longer particles were initially located to the top or the the bottom of the mixture. A proportionality between AR difference and segregation intensity was observed. Some variables such as rotational speed and initial bed configuration were found to not affect the segregation dynamics and intensity. Only filling level (in addition to AR) was found to affect segregation intensity. A mechanistic explanation was given in terms of differences in particle flowability occurring in the surface active layer (confirmed also by an analysis of the angles of repose in the drum). In this specific case therefore percolation and inertial effects (typical of size segregation) were not the main driving mechanisms for the observed segregation (the particle indeed had the same volume and similar mass so that sterical and inertial phenomena were minimized). Segregation mainly occurred because of differences in particle velocity in the active layer due to a reduced mobility of particles with larger AR.

References

1. GDR MiDi (2004), "On dense granular flows", The European Physical Journal E, 14, 4, 341–365.
2. Jarray A., Magnanimo V., Ramaioli M., Luding S. (2017) Scaling of wet granular flows in a rotating drum, EPJ Web of Conferences, 140,03078.
3. Volpato S., Canu, P., Santomaso A.C. (2017), "Simulation of free surface granular flows in tumblers", Advanced Powder Technology 28(3), 1028-1037.
4. Mellmann J. (2001), "The transverse motion of solids in rotating cylinders – forms of motion and transition behaviour", Powder Technology, 118, pp 251-270.
5. Windows-Yule C.R.K., Scheper B.J., van der Horn A.J., Hainsworth N., Saunders J., Parker D.J., Thornton A.R. (2016), "Understanding and exploiting competing segregation mechanisms in horizontally rotated granular media", New Journal of Physics, 18, 2, 023013, 2016.
6. Santomaso, A.C., Olivi, M., Canu, P. (2004), "Mechanisms of mixing of granular materials in drum mixers under rolling regime", Chemical Engineering Science, 59 (16), 3269-3280.
7. Kumar N., Magnanimo V., Ramaioli M., Luding S. (2015), "Tuning the bulk properties of bidisperse granular mixtures by small amount of fines", Powder Technology, 293, 94-112.
8. Ramaioli M., Pournin L., Liebling Th. M., (2005), "Numerical and experimental investigation of alignment and segregation of vibrated granular media composed of rods and spheres", Proceedings of the 5th International Conference on Micromechanics of Granular Media, 2, 1359-1363.
9. Santomaso, A.C., Canu, P. (2007), "Single particles properties vs. bulk flowability", Proceedings of 5th International Conference for Conveying and Handling of Particulate Solids.
10. Schneider, C. A., Rasband, W. S., Eliceiri, K. W. (2012), "NIH Image to ImageJ: 25 years of image analysis", Nature methods 9(7): 671-675.

507441 Generation of Particles with a Special Morphology By Desublimation of Copper Phthalocyanine

Abstract:

The morphology of submicron particles becomes more and more important concerning new functional materials. For example the special electric properties of particles can be used in electronic devices [1]. Moreover the electric, optical or catalytic properties of particles can be optimised by adjusting the morphology of the particles. This may imply a risk for the health but makes it possible to realize specific product characteristics. The morphology of the particles should therefore be considered e.g. for filter tests and for the developing and testing of sensor systems.

To explore the characterization of the particle morphology the DFG-AiF cluster project "multi-parametric characterization of particle-based functional materials using innovative online measurement systems" (MPaC) includes the development of online sensor systems for the size and structure of the particles as well as the development of generation methods for particles with a special morphology. The production of spherical particles can be achieved with different commercial aerosol generators. Several generation techniques for special particles are described in literature [2] [3] [4]. To test the sensor systems with particles of different shapes a new type of aerosol generator was developed. The aim for the new generator is to produce the particles in the gaseous phase and provide the particles directly for sensor or filter testing.

The developed generator uses the desublimation of copper phthalocyanine (organic pigment) to structure the surface of titanium dioxide. The bulk material copper phthalocyanine is dispersed into a nitrogen gas stream and transported to a tube furnace. Here the bulk material sublimates and the gas stream carries the material to a laval nozzle where the adiabatic expansion takes place. This leads to a rapid decrease of the temperature behind the laval nozzle and an oversaturation in the gaseous phase. In addition to the decrease of the temperature core particles of titanium dioxide were initiated behind the laval nozzle. The copper phthalocyanine desublimates on the core particles and causes a structured surface. The morphology of the generated particles is characterized by SEM. The particles show a punctual coating of the titanium dioxide particles with copper phthalocyanine. The size of the titanium dioxide particles is about 500 nm and the punctual coating of copper phthalocyanine measures a few nanometres.

The generation method and the design of the generator will be presented and the morphology of the generated particles will be described.

[1] K. Deppert, M. H. Magnusson, S. B. Carlsson, B. Gustafson, T. Junno, . L. Montelius, B. J. Ohlsson, L. Samuelson und L. E. Wernersson, „Future electronic devices by aerosol techniques," *Journal of Aerosol Science,* Nr. 31, 2000.

[2] J. Raula, A. Kuivanen, A. Lähde, H. Jiang, M. Antopolsky, J. Kansikas und E. . I. Kauppinen, „Synthesis of image-leucine nanoparticles via physical vapor deposition at varying saturation conditions," Journal of Aerosol Science, Nr. 38, 2007.

[3] M. Heurlin, M. H. Martin, D. Lindgren, M. Ek, R. L. Wallenberg, K. Deppert und L. Samuelson, „Continuous gas-phase synthesis of nanowires with tunable properties," Nature, Nr. 492, 2012.

[4] K. Nielsch, K. Deppert, M. H. Magnusson, G. Dumpich, E. Kruis und H. Fissan, „Formation of ultrafine particles from powders of compounds with incongruent sublimation behavior," Journal of Aerosol Science, Nr. 28, 1997.

516184 Rubbery Milling of Seed Endosperms for Improved Sustainability by Natural Functionality Preservation

Abstract:

Increasing demand for functionalised goods addressing individual needs requires a profound understanding of process-induced alterations. Development of models and process-structure-property relationships are essential to control the resulting quality and properties of processed food. A wide range of food systems contain solid or semi-solid particulates as major structure building blocks. Generation and tailoring of such particulates strongly depends on their size distribution, shape, morphology, phase behavior (amorphous/crystalline), interfacial properties and related inter-particle force interactions.

A novel processing technology was developed widening the frame of possible alterations, aiming towards tailoring such particle properties generating novel food structures or preserving techno-functional characteristics of the raw material. The technology employs targeted adjustments of the initial moisture content in combination with milling temperature. The resulting transition from the glassy to the rubbery state generates different material properties and processing behaviour. The technology is highly versatile, owing to the universality of moisture induced state transitions.

Seeds from locus bean gum or guar gum were wetted with excess water in order to adjust a rubbery state of the seed endosperm material. In this state milling within a novel type of impact mill with superimposed drying and sifting steps causes more gentle material disintegration compared to conventional dry milling. This preserves the high molecular weight fraction of e.g. galactomannans (micro-scale structure) from degrada-tion and allows at the same time the adjustment of a micro-porous morphology (macro-scale structure) improving wetting, water uptake and dispersing characteristics of the resulting thickener powders. Such structure preservation /modification increased the thickening characteristics extraordinarily as will be demonstrated.

From this prominent rubbery milling example we have entered in related systematic work addressing a wider range of seed endosperms with various â€œnature-synthesizedâ€ func-tionnalities to be further addressed.

/1/ Pollard, M., Eder, B., Fischer, P., and Windhab, E. J. (2010) "Characterization of galacto-mannans isolated from legume endosperms of Caesalpinioideae and Faboideae subfamilies by multidetection aqueous SEC." *Carbohydrate Polymers*, 79(1), 70-84.

/2/ Illmann, S., Stoob, M., Sieber, C., Konz, A.K. and Windhab, E.J. (2012) "New Thickening Quality of Galactomannan Polysaccharides by Tailored Impact Milling" *XVIth Int. Congress on Rheology*, Lisbon, Portugal, August 5-10

503422 Particle Technology Education at Purdue University

Abstract:

Degree programs and coursework at Purdue University focused on particle technology will be presented. Additional information may be found at: https://engineering.purdue.edu/CP3

506453 Teaching Particle Technology in Portugal â€“ University of Coimbra

Abstract:

Education in Particle Technology started in Portugal in the 80â€™s (last years of the decade). The Chemical Engineering degree of the University of Coimbra was the first engineering degree, in the country, to include this topic in the curriculum. Particle Technology was recognized by us, at that time, essential for the education of chemical engineers, in parallel to education in the traditional areas of Chemical Engineering which were still focused on monophasic systems.

Since then, Particle Technology has always been present in the Chemical Engineering curriculum of the University of Coimbra, in different configurations. During the first years particle characterization was combined with design of operations involving particulate systems (grinding and pneumatic and hydraulic transport). From 2005 to 2010, with the curriculum reformulation in relation with the Bologna Process, the two topics started to be addressed in different courses: one course on Materials Characterization and the process topic in a course entitled Structuring and Stabilization Processes, dealing with colloids and emulsions, grinding, agglomeration and spray drying and, for the first time, an integrated modelling strategy for particulate processes based on the population balance formulation was introduced in this course. The two courses were compulsory, at this time.

From the school year 2011/2012 till present, the configuration of this area in the Chemical Engineering degree of the University of Coimbra changed again, even if not deeply (see https://apps.uc.pt/courses/EN/course/361). The course on materials characterization, which includes general notions on particle characterization is kept as an autonomous course included in the 1st cycle of the Chemical Engineering degree of the University of Coimbra. A second course on design of processes involving particulate systems now exists in the 4th year of the MSc degree (1st year of the 2nd cycle), entitled Disperse Systems Technologies, which is an optional course, and which deals with all types of disperse systems, i.e gas, liquid or solids as disperse phase. This course starts with a review of the mechanisms/interactions occurring at the interface with the disperse phase, goes on, again, with a chapter on population balances modelling, and proceeds with strategies for disperse systems stabilization, linking this subject with formulation concepts, ending with the design of some processes involving particulate systems, such as colloids and emulsions, crystallization, grinding, agglomeration and spray drying. This course is also offered to other engineering degrees, such as Materials Engineering and Biomedical Engineering.

Also at the PhD level a course on Disperse Systems Technologies can also be offered during the formal lectures period of the PhD program on Chemical Engineering (or others), depending on the students background and interests.

Since the introduction of the Particle Technology topic in the curriculum of the Chemical Engineering degree of the University of Coimbra, the feedback from employers has been very positive, and the competences in this field acquired by our students have been quite valued by the job market. Till today, the Chemical Engineering degree of the University of Coimbra is still the one, in Portugal, with a stronger emphasis on this topic in the curriculum. In our curriculum, the topic of Particle Technology interacts directly with another course on Product Design, also in the 4th year of the MSc degree and, of course, also with the Project Design course (5th year of the MSc degree), besides other more traditional courses in a Chemical Engineering degree such as in Reaction Engineering and with courses related to environmental technologies.

At present, we have started developing multimedia tools for the teaching/learning of this topic, especially in the field of particle characterization, to be made available on-line, as is the case of the interactive video on nanoparticles characterization (http://labvirtual.eq.uc.pt/siteJoomla/index.php?option=com_content&task=view&id=277&Itemid=470, in Portuguese). Also, the European Federation of Chemical Engineering (EFCE) is considering investing in this type of tools for the field of particle technology, as the result of a collaboration between the Working Groups of Characterization of Particulate Systems and Education.

Acknowledgement

The authors would like to acknowledge the financial support of Pest/C/EQB/UI0102/2013 from FCT/MCTES (PIDDAC), co-financed by the European Regional Development Fund (ERDF) through the program COMPETE (POFC).

506629 Highlights of Particle Technology Teaching in Singapore

Abstract:

Our approach to education in particle technology is to incorporate elements of our research in this field, relevant industrial case studies and apply a Technology Enhanced Learning (TEL) approach for implementing pedagogical interventions to enhance the learning experiences of students.

For supercritical CO_2 processing, case studies in extraction and drying are used to teach different chapters in the curriculum such as:

- Particle characterization
  - The bulk density, size, shape of real industry samples are given to students for analysis using software such as ImageJ
- Flow through packed bed
  - The pressure drop and scale-up of packed bed design for supercritical CO_2 flow through packed bed are used as a worked example and assignment problem to help students appreciate the role of density and viscosity in the pressure drop in a packed bed
- Drying
  - Supercritical CO_2 as an alternative drying media as compared to hot air or superheated steam. Case studies and examples provided

For gasification and chemical looping combustion applications, case studies in the treatment of wastes and carbon capture are similarly used as part of the syllabus:

- Particle characterization
  - The effect of polydispersity on various fluidization characteristics (entrainment, transport dis-engaging height, clusters, segregation, etc)
  - The effect of agglomeration
- Reliability of correlations and simulations
  - Correlations in literature give predictions orders-of-magnitude apart
- Scale-up
  - Rule-of-thumbs in scaling up the various regimes

A design software has been developed within the paradigm of Technology Enhanced Learning (TEL) to facilitate learning via a repeated practice approach. The software generates detailed solution steps to typical engineering design problems encountered within this module. Students can compare the complete solutions generated with their own hand calculations, and thereby apply a repeated practice approach towards their learning of engineering design calculations. The software has been successful in enhancing the performance of students in the final examination.

507116 Particle Technology Course at the University of Salerno

Abstract:

The Course of Particle technology at the University of Salerno is a 6 EC, that is a 60 hours course including lectures, as well as classroom and laboratory practical. The course dedicated to Master student in Chemical Engineering. It a compulsory course within the â€œinnovative process and nanotechnologies curriculumâ€ can be optioned by any mechanical or chemical engineering student at the master level.

The course includes basic particle size and shape characterization concepts, fundamentals on interparticle forces, fundamentals of powder mechanics and silo discharge, fundamentals on-fluid particle interactions and fluidization.

The main learning outcome is the ability of the student to understand the phenomena occurring in processes involving particulate solids and to carry out basic design operation on storage and fluidized bed units.

507382 Professional Master of Engineering Degree in Particle Technology

Abstract:

Particles are ubiquitous in the process industries yet nearly absent in the undergraduate education of chemical engineers. This poster descibes a professional Master of Engineering degree program at University of Delaware (the MEPT) that closes that gap by exclusively focusing on particulate matter, its processing, and its incorporation into products. The program can be completed in a single year although we welcome already-employed students who wish to take it part time. Now in our fourth year, we will describe our program and our experiences including guiding distance learners successfully through the program.

In its 1-year realization, a core curriculum of 18 credits is augmented by 6 credits of electives which the student may use to tailor the program to their specific interests. The program is capped by a 10-week, 6-credit summer internship. The 12 core credits taken in the fall teach the fundamentals of particle transport, particle properties & characterization, particle formation and transformation kinetics and the mathematics of particulate systems. The 6 core credits taken in the spring teach the student to apply the fundamentals to both (a) equipment design (in a unit operations course) and (b) product and process design (in a product engineering course).

The presenters of this poster are the program directors, both with long industrial careers preceding this venture into education. In addition, we have developed a strong network of industrial partners who share a wealth of additional experience and expertise with our students both as adjunct professors and as internship mentors. We have successfully placed international students in internships at US companies via the curriculum practical training program (CPT) who have gone on to practice in industry after graduation under the optional practical training program (OPT).

507780 Recent Developments in Particle Technology at the Universidad Tecnica Federico Santa Maria

Abstract:

The Department of Mechanical Engineering of the Universidad Técnica Federico Santa María formed in 2003 a Research Center for Transporting Granular Materials (CITRAM) to support educational activities and experimental research, as an extension for those ME students taking the regular course on Bulk Solids Handling and those developing a thesis in this area.

Over the years, many ME students have studied, investigated and modeled different topics including: pneumatic conveying of solid particles of different nature, size and density; identification of flow patterns developed in the pipeline, minimum conveying velocity and pressure drop; abrasive wear using a novel rubber-wheel tester; segregation of ores in stockpiles and silos; flow of granular materials in hoppers and chutes; heat transfer of granular materials; among others. The aim of this poster is to present and share the results of this work.

508520 Using Perusall to Enhance Student Learning of Particle Technology at Graz University of Technology

Abstract:

Perusall is a digital platform that allows students to collaboratively annotate textbooks, papers, and other similar classroom reading assignments. In order to enhance student learning, Perusall was introduced to help teach Particle Technology at Graz University of Technology (TU Graz, Austria). At Graz University of Technology, two Particle Technology courses exist in sequence; the first course is focused on fundamentals topics, whereas the second course is focused more on applications and separations topics. Approximately 60-75 students are registered for each course. For both courses, Perusall was used to assign reading materials, which were primarily journal articles that complemented the course topics. Although these assignments were non-compulsory due to national requirements, students were asked to use Perusall to read and annotate the course readings. Student participation and input regarding this digital approach to learning was recorded. This poster will highlight how Perusall was implemented in both Particle Technology courses as a tool to enhance student learning.

508524 A Graduate Course in Fluidization and Gas-Solid Flow Systems

Abstract:

The objective of this course is: a) to provide students with a fundamental understanding and design of Fluidization and Fluid- Particle flow systems, b) to provide students with an understanding of Computational Fluid Dynamics (CFD) and the kinetic theory approach to fluid-particle flows c) to provide students with the knowledge required to solve flow, heat and mass transfer problems in processes which involve particles and d) to provide students with sufficient knowledge to read and understand technical articles in areas related to fluid-particle processes. The course is designed for Master and PhD students in Chemical, Mechanical and aerospace engineering. The course includes: Fluidization concepts and regimes including bubbling, turbulent and circulating beds; Particle characterization, diagnostics and size reduction and agglomeration/granulation; Concept and design of Standpipes, non- mechanical valves, gas distributors, cyclones and hoppers; CFD approach to fluid-particle flow systems and fluidization including kinetic theory and dispersed phase modeling; Computational transport phenomena of fluid-particle systems including method of moments; and examples from energy, chemical and pharmaceutical industries. In addition the students will visit PSRI to see large-scale fluidized beds in operation.

The students are required to finish a team project. The teams consist of 3-4 students working on a project which will count as their final exam. The projects will primarily involve background research (about 80%) but some of the investigation (about 20%) should involve taking an additional step from what is currently known. This additional step can take the form of a research proposal, new theory, new insight based on current information, new calculations, new simulation and/or new data. The results of the projects will be communicated to the class and to the Course Instructors via a written report of 10-20 pages, and an oral presentation involving the professors and all of the group members.

508742 Education of Fluid-Solid Multiphase Flow at Department of Mechanical Engineering, Osaka University

Abstract:

In mechanical engineering field, education curriculum in undergraduate course concentrates on the basic subjects of mechanical engineering such as mathematics, mechanics, material mechanics, fluid mechanics, thermodynamics, control engineering, design, etc. Concerning the course relating to powder technology, we have a class of "Fluid-Solid Multiphase Flow" in the curriculum of graduate school of engineering.

The contents of "fluid-solid two-phase flow" is as follows;

1 Introduction: examples of multiphase flow

2 Definitions and fundamental concepts

3 Particle response time and Stokes number

4 Fluid force acting on particle

5 Fluid force acting on particle in dense system

6 Particle-wall and particle-particle interaction

7 Heat and mass transfer

8 Continuous phase flow equations

9 Modeling of fluid-solid multiphase flow

10 Discrete particle modeling and simulation

11 Numerical simulation method for continuous phase flow

This course includes basic concepts of powder technology, dynamics of particles in fluid flow, derivation of basic equations for locally-volume-averaged fluid flow, discrete particle modeling including DEM (Discrete Element Method), and basics of numerical analysis and modeling.

509420 Bulk Solids Handling Education at the KSU Bulk Solids Innovation Center

Abstract:

Kansas State University Bulk Solids Innovation Center (BSIC) was started in 2015 with university and industry partnership and is located in Salina, Kansas. This center is dedicated to bulk solids handling research, education and outreach and equipped with world class pneumatic conveying system (both dense and dilute) full scale test lab and other material testing equipment. The vision of this center is to become valued resources for companies who produce or handle bulk solids during their operations. Currently, this center offers short courses in various subject lines related to bulk solids handling such as storage and flow, pneumatic conveying systems, discrete element method, dust explosion and control and mixing and segregation. Duration for these short courses varies from two to three and half days and offered as live course with class room teaching and practical training. The fundamental course on bulk solids handling is offered as distance education online format. This center is also in the process of starting two year online (distance education) master degree program named professional master of technology specialized in bulk solids handling and one year graduate certificate program (online) in bulk solids handling.

506839 Atomic Layer Deposition for the Synthesis of Nanostructured Catalysts

Abstract:

Atomic layer deposition (ALD) is a thin film growth technique that uses cycles of alternating, self-limiting chemical reactions between gaseous precursors and a solid surface to deposit material in an atomic layer-by-layer fashion. Saturation of the individual ALD surface reactions and gaseous diffusion of the chemical precursors ensure that all exposed surfaces of a template become uniformly coated. These attributes, coupled with a broad palate of available materials, make ALD a versatile technique for synthesizing catalysts. Conceptually, ALD can be used to build a catalyst as a series of discrete layers which perform specific functions such as tuning the porosity, serving as the catalyst support, providing catalytic activity, and imparting thermal stability. This capability allows the physical structure and porosity of a template to be decoupled from the surface chemistry and catalytic properties. Consequently, the same catalyst can be synthesized simultaneously on a variety of templates such as a planar single crystal to facilitate detailed characterization, or a high surface area powder to boost conversion for catalytic testing.

In addition to depositing continuous layers, ALD also enables the synthesis of nanoparticles. Using ALD, the nanoparticle size can be modulated by adjusting the number of cycles performed, and the number of nanoparticles can be controlled by pretreating the surface to change the concentration of surface functional groups on which the ALD initiates. By alternating between different ALD surface chemistries, mixed-metal nanoparticles (e.g. Pt-Pd) can be formed. Furthermore, the layer-by-layer nature of ALD enables specific nanostructures for these mixed-metal nanoparticles such as core/shell (Pd/Pt or Pt/Pd) and well-mixed alloy. Finally, ultrathin ALD metal oxide layers applied after the metal nanoparticle ALD can stabilize the nanoparticles to inhibit sintering at elevated temperatures. Quite remarkably, these metal oxide â€œovercoatsâ€ grow selectively on specific sites on the noble metals and on the support between the nanoparticles such that the catalytic activity is maintained or even enhanced.

509584 The R & D on Engineered Particles for Functional Materials in China

Abstract:

This is an overview on the R & D activities of functional powder processing from the particle level. China needs this kind of technologies to support its new material industry and improving the qualities of commonly used materials.

It starts from the national policy in this 10-year action plan designed to transform the structure of processing industry and increase the quality of â€œMade in Chinaâ€. The Key Developing Sectors are: Aeronautics/Astronautics, New Generation IT, Machine Tools and Robots, Marine Engineering, Advanced Rail Transportation, New Energy Vehicles, New Energy, and New Materials.

Particle coating and hybridization is an important approach to make functional powders. Newly developed equipment is introduced. Copper/alumina, PE/CNT composite particles are prepared. Particle shape modification is another technology to increase the powder flow-ability and higher tap-density. The technology is introduced with the analysis of the process on graphite powder process for Lithium-ion Battery Anode.

507968 Atomic Layer Deposited Nickel Nanoparticle Catalysts for Dry Reforming of Methane (Invited)

Abstract:

Highly dispersed nickel (Ni) nanoparticle catalysts were prepared by atomic layer deposition using bis(cyclopentadienyl)nickel and hydrogen as precursors at 300 °C. Both porous γ-Al_2O_3 particles and dense γ-Al_2O_3 nanoparticles were used as the substrates. For comparison, Ni nanoparticles supported on porous γ-Al_2O_3 particles were also prepared by the incipient wetness (IW) method using an aqueous solution of $Ni(NO_3)_2 \cdot 6H_2O$. The Ni/alumina catalyst was employed to catalyze the reaction of dry reforming of methane (DRM). It was found that the ALD Ni nanoparticles were much more active than the Ni nanoparticles synthesized by the IW method; the porous γ-Al_2O_3 supported ALD Ni nanoparticle catalyst showed higher activity and greater stability, as compared to the dense alumina supported ALD Ni nanoparticle catalyst. The porous γ-Al_2O_3 supported ALD catalyst showed exceptionally high catalytic activity (a methane reforming rate of 1840 $Lh^{-1}g_{Ni}^{-1}$ at 850 °C) and excellent stability of DRM reaction in over 300 hours at temperatures that ranged from 700 °C to 850 °C. The greater stability could have been due to the porous structure of the support and the formation of $NiAl_2O_4$ spinel during the ALD coating process, which could limit the sintering of the Ni nanoparticles and, thus, inhibit coking during the reaction. The high catalytic activity was due to the high dispersion of Ni nanoparticles deposited by ALD and the reduction of $NiAl_2O_4$ spinel to Ni during the DRM reaction at 850 °C. It was verified that $NiAl_2O_4$ can be reduced to Ni in a reductive gas mixture (i.e., carbon monoxide and hydrogen) during the reaction at 850 °C, but not by H_2 alone.

507836 Extended Surface Electrocatalysts Synthesized Via Atomic Layer Deposition

Abstract:

Platinum-nickel nanowires are a promising advanced catalyst material for the oxygen reduction reaction (ORR) in fuel cell applications. When synthesized by galvanic displacement of nickel nanowires (NiNWs), these extended surface electrocatalysts have demonstrated activity and durability superior to benchmark carbon-supported Pt nanoparticle catalysts. However, challenges in scale-up and reproducibility of this fabrication process have prompted investigation into alternative synthesis routes. In this work, atomic layer deposition (ALD) of Pt was used to develop a scalable analog process to the galvanic displacement of extended surface electrocatalyts. Pt was deposited on nickel nanowire substrates in packed bed reactor equipped with in situ mass spectrometry to monitor products of the ALD half-reactions. The ALD process was investigated using both O_2 and H_2 as ligand removal agents. Resulting PtNi nanowire catalysts were characterized at 5, 10, 20, and 30 cycles of ALD using a variety of analytical techniques (chemisorption, ICP-MS, electron microscopy) to analyze the nucleation and growth of the Pt during the deposition process and its effects on active surface area. In the case of the oxidative ALD process, these novel extended surface catalysts were found to possess an ORR mass activity almost 3X higher than benchmark Pt/C and over 70% retention of activity after durability testing.

508418 Scalable Manufacturing of Nanostructured Noble-Metal Catalysts Using Atomic Layer Deposition

Abstract:

Nanostructured particles are extremely useful in applications such as catalysis, energy storage and pharma. However, a hurdle in their utilisation is that typically large amounts of such materials are required. Current liquid-phase and gas-phase synthesis methods often lack the high precision required or do not lend themselves to large-scale production. Gas-phase coating using atomic layer deposition (ALD, a variant of chemical vapour deposition) can be used to provide the surface of a particle with either an ultrathin continuous coating or a decoration of nanoclusters. When carried out in a fluidized bed, ALD is an attractive way of producing nanostructured particles with excellent scale-up potential.

Yet, exploiting the precision of ALD to have full control over the production of catalytic materials not straightforward. The precision of ALD relies, in fact, on cyclic repetitions of self-saturating surface reactions that lead to the deposition of less than a monolayer per cycle. Hence, if the growth proceeds in a layer-by-layer fashion, as is the case with ALD of thin films, varying the number of cycles translates into tuning the film thickness with atomic-level precision. However, if the as-deposited atoms form into NPs upon deposition, a growth per cycle of less than a monolayer does not necessarily enable atomic-level control over the NP size. This is because the NP morphology (e.g., size, shape and number density) is dictated by atomistic processes other than "ALD reactions" such as surface diffusion and aggregation of atoms and NPs, and atom attachment to and from NPs. Understanding the role of such kinetic processes during the ALD of NPs is therefore crucial to the development of ALD routes for the synthesis of NPs with a well-defined morphology and thus functionality.

Here, we present an atomistic understanding of thermal ALD of Pt and Pd NPs on oxides nanoparticles and graphene nanoplatelets. In particular, we study the effect of temperature, number of cycles, coreactant partial pressure and exposure time on the evolution of the NP size distribution and metal loading. Atomistic modelling of our experiments shows that the NPs grow mostly due to Smoluchowski aggregation, that is, NP diffusion and coalescence, rather than through single atom processes such as precursor chemisorption, single atom diffusion and attachment, and Ostwald ripening [1]. While the metal loading can be precisely controlled over a wide range of temperatures, atomic-level precision over the NP size is retained only at low deposition temperatures ($T \leq 100$ °C) when growth by atom attachment/deposition becomes relevant. Furthermore, we show that the coreactant partial pressure and exposure time are far more important parameters for the NP size than the number of cycles. We will demonstrate the favourable effect of optimizing the synthesis conditions on the catalytic performance of the obtained material.

[1] Grillo, Fabio, et al. "Understanding and Controlling the Aggregative Growth of Platinum Nanoparticles in Atomic Layer Deposition: An Avenue to Size Selection." The Journal of Physical Chemistry Letters 8.5 (2017): 975.

509588 Flame-Made TiO_2(B)

Abstract:

TiO_2 is the second largest commodity made by flame reactors, outnumbered in tonnage only by carbon black. Unless dopants are employed to increase the rutile yield, e.g. for application as white pigments, product powders are a mixture of anatase and rutile polymorphs. One prominent titania nanopowder is high surface area (~50 m^2/g) titania P25 (Evonik Industries AG) with approximately 88 wt% anatase and 12 wt% rutile [1] that has broad application as a photocatalyst.

However, TiO_2 can also attain other crystal phases than anatase and rutile, namely brookite, TiO_2â…¡, TiO_2(H), and monoclinic TiO_2(B). The latter has the lowest density and most open crystal structure of all titantia polymorphs, while its energy bands differ from and interact with the energy bands of the other crystal structures [2]. Consequently, it may lead to photocatalytic activity similar to or better than P25 [3].

Here, titania mixtures with monoclinic (TiO_2(B)) and anatase crystal phases are synthesized using flame spray pyrolysis (FSP). X-ray diffraction, transmission electron microscopy, and Raman spectroscopy confirmed the presence of TiO_2(B). Raman measurements detected up to 27% of TiO_2(B) in the crystalline fraction. The amount of TiO_2(B) increases as the time for nanoparticle growth in the flame decreases through quenching by mixing with ambient air. It can be controlled by varying the flow rates of precursor solution and dispersion oxygen into the flame. Those FSP-made titania outperformed commercial titania (P25) in the photocuring of epoxy.

1) Ohtani B.; Prieto-Mahaney, O. O.; Li, D.; Abe, R. What is Degussa (Evonik) P25? Crystalline Composition Analysis, Reconstruction from Isolated Pure Particles and Photocatalytic Activity Test. *J. Photochem. Photobiol., A* **2010**, 216, 179-182.

2) Armstrong, A. R.; Armstrong, G.; Canales, J.; Bruce, P. G. TiO_2-B Nanowires. *Angew. Chem., Int. Ed.* **2004**, 43, 2286-2288.

3) Yang, D.; Liu, H.; Zheng, Z.; Yuan, Y.; Zhao, J.; Waclawik, E. R.; Ke, X.; Zhu, H. An Efficient Photocatalyst Structure: TiO_2(B) Nanofibers with A Shell of Anatase Nanocrystals. *J. Am. Chem. Soc.* **2009**, 131, 17885-17893.

507815 Dopant Modified Iron Based Oxygen Carriers for Chemical Looping Combustion and Gasification Applications (Invited)

Abstract:

Chemical looping technology is promising in high efficiency clean carbonaceous fuel conversion, including chemical looping gasification and chemical looping combustion. The reactions in chemical looping involve cyclic redox reactions with metal oxide as the oxygen carriers to provide lattice oxygen and regenerate in air. Iron-based oxygen carriers modified with both isovalent and aliovalent transition metal ions have been investigated in this work. Isovalent dopants can lower fuel conversion reaction energy barriers by providing active sites for fuel activation and conversion during redox reactions whereas aliovalent dopants can not only provide such active sites but also create Schottky defects to increase the ionic conductivity in the oxygen carriers. In this study, we examine the impact of both isovalent and aliovalent dopants on iron oxide oxygen carriers reactivity and conversion rates using methane reforming and CO combustion as examples. A screen of suitable transition metal ion dopants have been done by DFT calculation as well as experiments. A detailed characterization and reaction mechanism will be discussed. The understanding of the reaction mechanism provides insight into dopant modification for improving the reactivity and recyclability of oxygen carriers, which is essential in commercialization of the oxygen carriers for chemical looping gasification and combustion.

508069 Porous Composites As Host Materials for Lithium-Sulfur-Batteries (Invited)

Abstract:

Scalability, efficiency and economic feasibility are important aspects for the fabrication of cathode materials for lithium-sulfur batteries. For the application of lithium-sulfur batteries in electric vehicles and stationary energy storage systems, different challenges need to be overcome to achieve sufficient energy and power density within weight and volume constraints. Further safety requirements and the reproducible fabrication of large active material quantities at low costs have also to be guaranteed for a wide range of applications [1]. Sulfur allows high theoretical capacities of 1675 mA h g^{-1} as cathode material in comparison to lithium, resulting in a theoretical energy density of around 2600 W h kg^{-1} for lithium-sulfur batteries [2-5]. However, using sulfur as active material leads to compositional and structural changes and large volumetric expansion up to 80 % during cycling, aspects which must undoubtedly be considered. Besides, reducing the formation of polysulfides as adverse byproducts and insulating Li_2S is still a challenge to be solved [1, 6, 7]. As a consequence of these insufficient cycle stabilities, low coulomb efficiencies [3] and high self-discharge rates [4, 8], a barrier has been built against commercialization [2, 7, 9-11]. Porous host materials with defined pore structures which are able to incorporate certain contents of sulfur have been postulated as a possible solution to compensate the drastic volumetric expansion during cycling. A widely-used method to incorporate sulfur within host materials is the melt-diffusion process.

As published in Zellmer and Titscher et al. [12], different aspects, such as the used host material and its specific properties, the parameters of the comminution process as well as the formulation strategy to prepare the electrodes were investigated to evaluate their influence on the electrochemical performance of the lithium-sulfur-battery. Concerning the host material, the content of sulfur, the pore volume, the particle size distribution, the crystallinity and the conductivity of the material were characterized. Subsequently, the host materials, prepared via hydrothermal carbonization or purchased, were used for the comminution process in different mills. As published in [13 -16] planetary ball mills can be used to incorporate sulfur into host materials. The process can be optimized by adjusting the process time and including thermal treatments. Based on the specific energy input, sulfur can be embedded within the material but at the same time pores within the material can be removed [13-16]. Experimental studies and DEM simulation were performed to characterize the operation of different mills with regard to the incorporation of sulfur and to manufacture homogeneous composites without the combustion of filled pores. A planetary ball mill was chosen in this study in comparison to the literature and additionally, a vibration mill was used to prepare the sulfur carbon composites under lower stress energies than the ones arisen during the process in the planetary ball mill. The vibration mill enables the preservation of the pores while assuring simultaneous homogenization of the product and shows considerable potential for the scale-up [17-19]. For the evaluation of the influence of the mechanical treatment in the vibration mill on the preparation of homogeneous composites, different carbon host materials with various pore size distributions were processed and characterized after each process step regarding the content of incorporated sulfur and the resulting particle size and pore size distribution. Different aspects were found in comparison to the investigations dealing with the planetary ball mill. On the one hand, smaller particle sizes and thus, lower contents of incorporated sulfur within the carbon host material were detected due to the higher specific energy within the planetary ball mill. On the other hand, more specific power is transferred into friction events, and thus heat can be dissipated leading to an improved incorporation of sulfur in the pore structure. This fact can also be discerned and determined via DEM simulations. Further on, only a homogenization and not an elimination of the composite (sulfur and host material) take place because of the lower frequency of events in normal direction [20]. Subsequently, electrodes were produced in order to characterize the material, understand the fabrication process and determine the electrochemical performance of the manufactured composites. To this aim, water was used as environment-friendly solvent and PEDOT:PSS as conductive binder to fabricate cathodes for such lithium sulfur batteries [6, 21]. The influence of the composite structure after the manufacturing within the different mills was also discerned via the electrochemical characterization. After analyzing the electrochemical performance, it was clearly observed that higher specific capacities at low cycle numbers were reached using the composites prepared via the planetary ball mill process. Nevertheless, a constant decrease of the specific capacity was also detected over the lifetime. In the case of the composites manufactured in the vibration mill, lower specific capacities for the first cycles were measured, but constant capacities over the whole lifetime were achieved. Additionally, significant influences of the different host materials on the electrochemical performance were also investigated.

Various particle sizes and thus, different formulation strategies and electrochemical performances result from the specific energy input based on the different comminution devices, namely vibration and planetary ball mill. In current studies, the focus is set on the usage of a continuous process (vertical stirred media mill) starting with the homogenization of different porous, conductive host materials combined with sulfur as active material. Furthermore, the addition of different solid and liquid electrolytes can be experimentally fulfilled in the mill. Besides, DEM simulations were coupled with the experiments in order to support the outcomes, compare previous investigations and give insight into the process.

References

[1] A. Manthiram, Y. Fu, Y.-S. Su, Accounts of Chemical Research 46(5) (2012) 1125-1134.

[2] X. Ji, K.T. Lee, L.F. Nazar, Nature Materials 8 (2009) 500-506.

[3] J. Gao, M.A. Lowe, Y. Kiya, H.D. Abruña, Journal of Physical Chemistry C 115 (2011) 25132-25137.

[4] B. Guo, T. Ben, Z. Bi, G.M. Veith, X.-G. Sun, S. Qiu, S. Dai, Chemical Communication 49 (2013) 4905-4907.

[5] N. Jayaprakash, J. Shen, S.S. Moganty, A. Corona, L.A. Archer, Angewandte Chemie 123 (2011) 6026-6030.

[6] Y. Yang, G. Yu, J.J. Cha, H. Wu, M. Vosgueritchian, Y. Yao, Z. Bao, Y. Cui, ACS Nano 5(11) (2011) 9187-9193.

[7] Y.V. Mikhaylik, J.R. Akridge, Journal of Electrochemical Society 151 (2004) A1969.

[8] V.S. Kolosnitsyn, E.V. Kuzmina, E.V. Karaseva, Journal of Power Sources 274 (2015) 203-210.

[9] L. Chen, L.L. Shaw, Journal of Power Sources 267 (2014) 770-783.

[10] S.S. Zhang,, Journal of Power Sources 231 (2013) 153-162.

[11] Y.M. Lee, N.S. Choi, J.H. Park, J.K. Park, Journal of Power Sources 119 (2003) 964-972.

[12] S. Zellmer, P. Titscher, E. Wienken, A. Kwade, G. Garnweitner, Energy Storage Materials (2017), 9, 70-77

[13] J. Xu, J. Shui, J. Wang, M. Wang, H.-K. Liu, S.X. Dou, I.-Y. Jeon, J.-M. Seo, J.-B. Baek, L. Dai, ACS Nano 8 (2014) 10920-10930.

[14] M.R. Kaiser, J. Wang, X. Liang, H.-K. Liu, S.-X. Dou, 9th International Conference on Lead-Acid Batteries â€“ LABAT 2014 279 (2015) 231-237.

[15] M. Nagao, A. Hayashi, M. Tatsumisago, T. Ichinose, T. Ozaki, Y. Togawa, S. Mori, Journal of Power Sources 274 (2015) 471-476.

[16] Y. Cao, X. Li, I.A. Aksay, J. Lemmon, Z. Nie, Z. Yang, J. Liu, Physical Chemistry Chemical Physics 13 (2011) 7660-7665.

[17] T. Yokoyama, K. Tamura, H. Usui, G. Jimbo, International Journal of Mineral Processing 44-45 (1996) 413-424

[18] T. Inoue, O. K., International Journal of Mineral Processing 44-45 (1996) 425-435.

[19] H. Huang, J. Pan, P.G. McCormick, Materials Science and Engineering A 232 (1997) 55-62.

[20] S. Zellmer, P. Titscher, C. F. Burmeister, A. Kwade, G. Garnweitner, (2017) Fraunhofer Verlag, Produktgestaltung in der Partikeltechnologie, Band 8, 65-77

[21] T. Stöcker, A. Köhler, R. Moos, Journal of Polymer Science Part B: Polymer Physics 50 (2012) 976-983.

509516 Supported Nickel and Iron Oxide Nanoparticles for Catalytic Asphaltene Decomposition Under an Air/Vapor Atmosphere for Enhanced Oil Recovery (Invited)

Abstract:

The production of heavy and extra-heavy crude oils has increased recently due to the current energy demand and the depletion of conventional resources. However, these type of crude oils presents different problems related to its production, transporting and refining. Accordingly, several in-situ techniques have been developed for enhancing HO and EHO recovery and improve their properties. In this sense, nanoparticles have proven to work well in improving the physical and chemical properties of this type of oils, such as viscosity, API gravity, and content to heavy components such as asphaltenes.

This work aims to synthesize, characterize and evaluate for the first time supported nickel and iron oxide nanoparticles for catalytic asphaltene decomposition under an air/vapor atmosphere. Silica nanoparticles of 11 nm in size were synthesized by the sol-gel method and were characterized by size, SEM-EDX, surface area, FTIR, point of zero charge and thermogravimetric analyses (TGA). Metal oxide nanoparticles were synthesized over Silica surface following the incipient wetness technique using commercial reagents and solutions from laterite leaching. The extraction of Nickel-Iron consisted of a laterite leaching process by the use of solutions of sulfuric acid at various concentrations and the use of temperature. Which allowed as a final product a solution with these two components that were quantified using absorption spectrometry atomic, which yielded a majority of iron and its respective proportion in nickel according to the acid used. The particle size of nickel oxide and Iron oxide nanoparticles was estimated in 2 and 6 nm with 5.0 and 32.6% of dispersion over the support surface. Batch adsorption experiments performed for obtaining asphaltene adsorption isotherms for the support and the functionalized nanoparticles. It was observed that functionalized nanoparticles had higher adsorption capacity than the support. Further, TGA-FTIR experiments were performed for the asphaltenes in the absence and presence of the different nanoparticles evaluated.

A fixed asphaltene loading of 0.2 m^2/g was employed. Experiments conducted for an air flow of 100 mL/min up to 800°C. $H_2O_{(g)}$ was introduced to the system at a flow rate of 6.3 mL/min using a saturator filled with distilled water. Results demonstrated the enhancing of asphaltenes decomposition by the action of the nanoparticles due to the reduction in the decomposition temperature of the asphaltenes up to 200°C in comparison with the system in the absence of nanoparticles. This work opens a wider landscape about the use of functionalized nanoparticles for improving the efficiency thermal enhanced oil recovery processes in the presence of air and vapor.

507640 A Novel Thin Film Deposition Method of Doped TiO2 Particles for Environmental Applications (Invited)

Abstract:

Nanoparticles are increasingly used for many applications targeting environment problems. The key challenge is its handling and performance. The latter is affected due to secondary processes such as aggregation and the former arise in transportation and safe disposal. This is solved by depositing the nano particle on a substrate, which are commonly termed as thin films. However the performance of thin films is limited to various physical factors, such as film thickness, configuration, material property etc which is often linked to the deposition method itself.

In this paper we present a novel deposition method named as chemical beam vapor deposition which allows the precise control of film growth rate and film configuration. The deposited film is charactirized and tested for its environmental application.

507972 Synthesis of Surface Functionalized Adsorbents for Adsorption of Metal Ions and Organic Pollutants By Atomic Layer Deposition (Invited)

Abstract:

Different thicknesses of TiO_2 and Al_2O_3 films were applied on micron-sized silica gel particles and biochars by atomic layer deposition (ALD), respectively. TiO_2 coated silica gel particles were used as adsorbents for adsorption of metal and metalloid ions; their adsorption ability for a mixture of 19 trace elements of heavy metals and other toxic elements, including As(V), Se (IV), Be(II), Al(III), V(V), Cr(III), Mn(II), Co(II), Ni(II), Cu(II), Zn(II), Ba(II), Tl(I), Sb(III), Cd(II), Ag(I), Sr(II), Mo (VI), and Pb(II), from aqueous solutions was investigated. The removal efficiencies of As(V), Se(IV), V(V), Mo(VI), and Pb(II) were higher than 75% simultaneously in the mixed solution at pH 3 for 20 and 40 cycles of TiO_2 coated silica gel particles without heat treatment. In contrast, the uncoated silica gel particles did not adsorb any ions under this pH condition. Al_2O_3 deposited biochars were used for removal of organic pollutants and the adsorption capacities were evaluated by adsorption of methylene blue (MB). The data fit well with the Langmuir isotherm and the maximum adsorption capacities were found to be 26.8 and 35.0 mg/g at 25 °C for the uncoated biochar and 5 cycles Al_2O_3 (5Al) coated biochar, respectively. The adsorbed MB amount per square meter achieved 1.3 mg/m^2 onto the 5Al biochar, and it was twice the amount on the uncoated biochar. The improvement of adsorption capacity after Al_2O_3 ALD coating was due to its improved hydrophilia.

Effect of Surface Modification on Filtration Performance of Gas-Liquid Coalescing Filters

Feng Chen, Zhongli Ji, Qiangqiang Qi*
Beijing Key Laboratory of Process Fluid Filtration and Separation, College of Mechanical and Transportation Engineering, China University of Petroleum, Beijing, PR China

Abstract

Great attention has been paid to the purification of liquid particles from exhaust streams by fibrous filters which are mostly wettable. However, it seems that non-wettable filters are more advantageous under certain conditions, with superior gas-liquid filtration performance. Hence, the requirement of modifying the surface properties of present wettable filters is increasing, because the fiber surface characteristics can not only alter shapes of liquid particles on the fiber surface, but also influence capture mechanisms of liquid particles carried by the airflow. Especially, the processes of coalescing, transporting and draining of liquid particles are closely related to the fiber surface properties.

The work aimed at experimentally investigating the effect of surface modification on separation of liquid particles by filters from air streams. Three different methods were applied to modify surface properties of glass microfiber filters, which were commonly in use in industries and were selected as the experimental control, being transferred from wettable to non-wettable. First method was using the silicon dioxide (SiO_2) nanoparticles, with a mean diameter of about 200 nm (greatly smaller than the fiber diameter), and they were physically grafted onto fiber surface. The second method was treating filters with fluoropolymer to reduce the surface energy of fibers. The third treating method combined the two strategies above, viz., nano coating and polymer fluorination. Therefore, the fibers possessing the best liquid-repellent performance were obtained. All filters were tested with an employment of a filtration performance experimental setup, aiming to evaluate the pressure drop, filtration efficiency and quality factor. The filtration efficiency was derived from both upstream and downstream particle number concentrations, simultaneously utilizing a scanning mobility particle sizer (SMPS) and an aerodynamic particle sizer (APS). The particle sizes taken in this study range from 0.05 to 20 microns.

It was illustrated in the results that the surface modification played an important role in the surface characteristics and filtration performance of fibrous filters. The three methods mentioned above were all able to change the wettability of filters in varying degrees. The average oil contact angles were 52°, 101°, and 121°, respectively, for these three different kinds of modified filters. It was found that the nanoparticles were uniformly distributed on the fiber surface by microscopic observation. Furthermore, a pronounced increase of equilibrium efficiency was shown from most modified filters, compared to the control. As expected, it was more difficult for the modified filters to be wetted by liquid, which resulted in a reduced equilibrium saturation, suggesting a longer operation lifetime for these filters. Meanwhile, the results also showed that all the modified filters presented a similar steady-state pressure drop as the control filter, indicating little effect of surface modification on the steady-state pressure drop of filters. On the whole, a filter, modified by combination of grafting nanoparticles and low-surface-energy polymer fluorination, had the overall best filtration performance in the steady state, with a highest quality factor of 0.69 kPa^{-1} for the droplets measured by the SMPS and of 2.02 kPa^{-1} for the droplets measured by the APS.

Keywords: Nanotechnology; Particle technology; Separations

1. Introduction

Gas-liquid coalescence filters are widely used in many industrial processes including long-distance natural gas transportation pipelines, compressor dry gas seal systems, metal and mechanical cutting processes, crankcase ventilation systems, compressed air purification, and so on, which are mainly used to remove liquid particles from air streams or oil mists [1]. These filters are commonly composed of fibrous materials, such as glass fibers, due to their high coalescence filtration performance [2]. It appears that most glass fibrous filters used in actual applications are wettable (or oleophilic), although non-wettable (or oleophobic) filters have obtained applications in some cases. Moreover, oleophobic filters are more advantageous in certain conditions due to less liquid retained in such filters resulting from low surface energy. Generally, the liquid retained in the filters will result in a reduced efficiency and increased pressure drop. Therefore, oleophobic filters may theoretically have a better filtration performance, which needs more investigation.

Research relevant with oleophobic filters mostly focus on three aspects, which are single fiber, sandwiched filter materials, and filter cartridge. At single fiber level, some work have been done on the interactive forces between fibers and droplets, droplet shapes on the fibers, and motions of droplets along fiber surface [3,4]. At filter material level, Kampa et al. [5,6] proposed a "jump-and-channel" model by experiments conducted with glass fiber filter media, and the effects of oil viscosity and filter properties on the pressure drop and liquid distribution were also analyzed. At filter cartridge level, the influence of oil loading rate and drainage layer on the filtration performance of oleophobic filters have been conducted by Chang et al. [7,8]. Additionally, two other work have been done on the effect of order of media wettability using a combined filter medium of wettable and nonwettable filters [9,10].

However, there are few works on the systematical analysis of gas-liquid filtration performance of filters with similar structures but different surface wettabilities. Best to our knowledge, there are only two relevant works. Letts et al. [11] investigated the effect of fiber surface wettability on pressure drop evolution and equilibrium classical filtration efficiency of three different filter materials. Their results showed that the high-surface-energy filter had a lower steady-state pressure drop due to less liquid hold-up, and initial pressure drop also differed among different filter materials. Another work was our previously published literature [12], in which we found that a superoleophobic filter that was modified using wet coating method presented a better filtration performance than untreated glass microfiber filter. Hence, surface modification is a useful method to prepare filters with unchanged structures but various surface energy, so that the effect of surface energy of filter medium can be analyzed. Nevertheless, there are many different methods to modify the filters, including addition of nanoparticles, fluorination, and combination of these two methods. Consequently, there is a growing need to analyze the effect of modification method on coalescence filtration performance of the filter. Also, this work may be of great potential meaning to the application and development of surface modification in the field of gas-liquid coalescence filtration.

The intent of this work is to analyze the effect of surface modification method on the filtration performance of glass fiber filters in terms of pressure drop, filtration efficiency, and quality factor, to study the liquid saturation of filters in the steady state, and to find an optimal treating method to obtain a filter with overall best filtration performance.

2. Experimental

2.1. Materials

The basic filter material used in this work was oleophilic glass fiber medium, of which the properties are listed in Table 1. Mean pore size was measured using a capillary flow pore size analyzer (CFP-1200AEF, Porous Materials, Inc.). Thickness was tested with a digital caliper. Grammage was obtained through weighing a filter piece with specific surface area using an electronic analytical balance (AL204-IC, Mettler Toledo). DEHS contact angle of filter material was measured using an optical tensiometer (Attension Theta, BiolinScientific).

Table 1. Properties of experimental glass-fiber filter materials

Filter material	Mean pore size (μm)	DEHS contact angle (°)	Thickness (mm)	Grammage (g/m^2)	Packing density (-)
GF	4.53±0.02	44.0±4.1	0.56±0.01	102±3	0.070±0.002

All the filters were composed of two same layers, including four kinds of filter materials. Based on surface treating methods of non-treatment, nanoparticle-grafting, polymer fluorination, and combination of above two, the filters were termed filter GF, GF-P, GF-F, and GF-PF, respectively. Filter GF was regarded as the control filter. Following section is detailed modification procedures.

(1) Preparation of nanoparticles: 1.0 g silica nanoparticles were weighed and dispersed into 100 ml distilled water by stirring 3 h. (2) Preparation of coating solution: 5.0 g perfluoroalkyl acrylic copolymer was dissolved in 100 ml distilled water, then stirred by a stirring bar for 30 min to form 5 wt. % milky emulation solution. (3) Preparation of coating solution with nanoparticles: 1.0 g silica nanoparticles were added into 5.0 g perfluoroalkyl acrylic copolymer with magnetic stirring overnight. Fully dispersed the as-prepared slurry in 100 ml distilled water by stirring. (4) Coating treatment: The glass fiber filter was immersed in the coating solution for 10 min to ensure that all the fibers were completely wetted. Then the filter was taken out from the coating solution, and excess liquid was dripped in fume-hood. Finally, the filter was heated in an oven for 1 h.

2.2. Experimental set-up and procedure

Fig. 1 shows the schematic diagram of filtration performance experimental set-up. The oil used in this work was DEHS, with density of 912 kg/m^3 and dynamic viscosity of 0.023 Pa s at 25℃. A six-jet aerosol generator (9306A, TSI) was used to produce the aerosol using compressed air purified by a HEPA (high efficiency particulate air) filter. The generated aerosol was mixed with clean air in a mixing chamber. The test filter whose efficient filtration area was 0.15 m in diameter was perpendicularly mounted in the filter chamber. The filtered gas was pumped using a vacuum pump after it was further purified by a HEPA filter. A mass flow controller was placed upstream the pump to control the face velocity during all experiments at 0.12 m/s. The captured liquid drained into a liquid collecting vessel under the force of gravity and air dragging force.

A scanning mobility particle sizer (SMPS, TSI) and aerodynamic particle sizer (APS, TSI) were used to simultaneously measure the upstream and downstream aerosol concentrations. By combination of these two instruments, droplets with diameters between 0.05 and 20 μm can be precisely analyzed. A pressure transmitter was used to continually measure and record the pressure drop of test filter. The measuring range and accuracy of each measuring equipment are listed in Table 2.

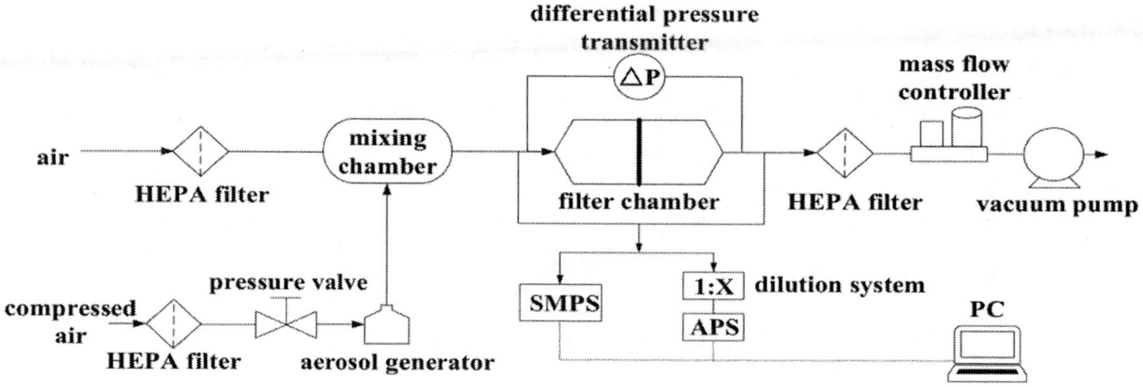

Fig. 1. Schematic diagram of filtration performance experimental set-up.

Table 2. Parameters of measuring equipment

Device	Model	Accuracy	Range
Mass flow controller	MCR 500 slpm	±0.4%	0–500 slpm
Differential pressure transmitter	EJX-110A	±0.04%	0–20 kPa
Electronic analytical balance	AL204-IC	0.1 mg	0–220 g
SMPS	TSI 3936	±4%	0.05–0.8 μm
APS	TSI 3321	±4%	0.8–20 μm

The filtration performance of filter is mainly evaluated in terms of pressure drop and filtration efficiency. To compare the overall filtration performance of all the filters, a quality factor (Q_F) was introduced [13] and defined by

$$Q_F = \frac{-\ln(C_{out}/C_{in})}{\Delta P} \tag{1}$$

where C_{in} and C_{out} are the inlet and outlet droplets number concentrations, respectively, and ΔP is the pressure drop across the filter.

3. Results and discussion

3.1. Characterization

Fig. 2a-d shows the scanning electron microscope (SEM) images of filter GF, GF-P, GF-F, GF-PF. As shown, nanoparticles were evenly distributed on the fiber surface of filter GF-P and GF-PF. A magnified image of deposited nanoparticles on filter GF-PF is presented in Fig. 2e, and it seems that there existed an obvious agglomeration phenomena among nanoparticles, which needs a further work to be prevented and improved. It can be found that filter GF-F that was treated with polymer fluorination shows no perceptible difference from untreated sample. Fig. 2f shows the results of pore size of different filters. The pore size distribution and mean pore size were almost the same among all filters as shown in Fig. 2f, suggesting that there is little influence of fiber surface modification on the pore size characterization of filters.

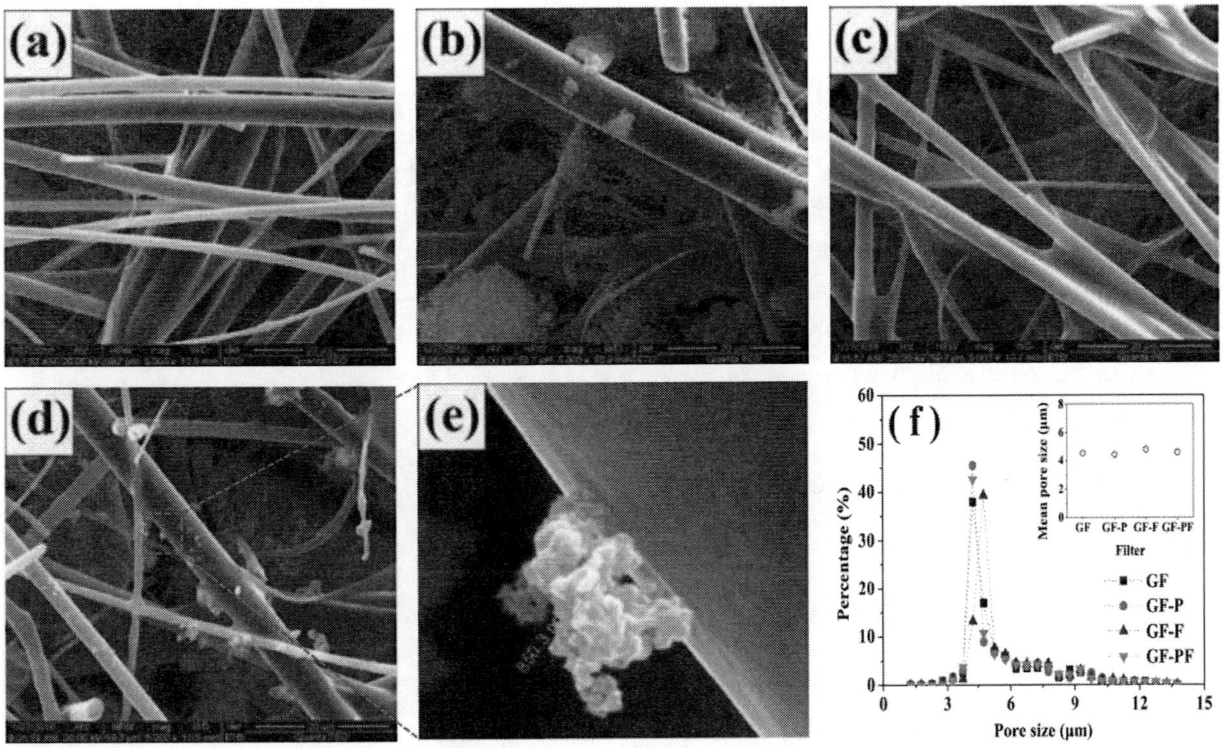

Fig. 2. SEM images of fibrous filter (a) GF, (b) GF-P, (c) GF-F, (d) GF-PF, and (e) magnified nanoparticles treated on the fiber of filter GF-PF; (f) pore size distribution of four filter materials.

Fig. 3 shows the DEHS contact angle and EDX result of different filters. As shown in Fig. 3a, the oil contact angle was improved for three modified filters when compared to the control filter. Filter GF-PF had a largest oil contact angle of 121.3°. Particularly, filter GF-F and GF-PF show an obvious oleophobic characteristic with average oil contact angles larger than 90°, indicating that the fiber wettability can be significantly changed through low-surface-energy fluoropolymer and addition of nanoparticles would further improve the oleophobic property of the filter. Fig. 3b shows that the increase of oil contact angle for filter GF-F and GF-PF was indeed related to the successful addition of element F on the fiber surface, due to relatively large amount of element F for these two filters while absent for other two filters. Besides, the contents of element Si and O appeared to greatly increase for filter GF-P and GF-PF after grafting nanoparticles on the fiber surface, as shown in Fig. 3b.

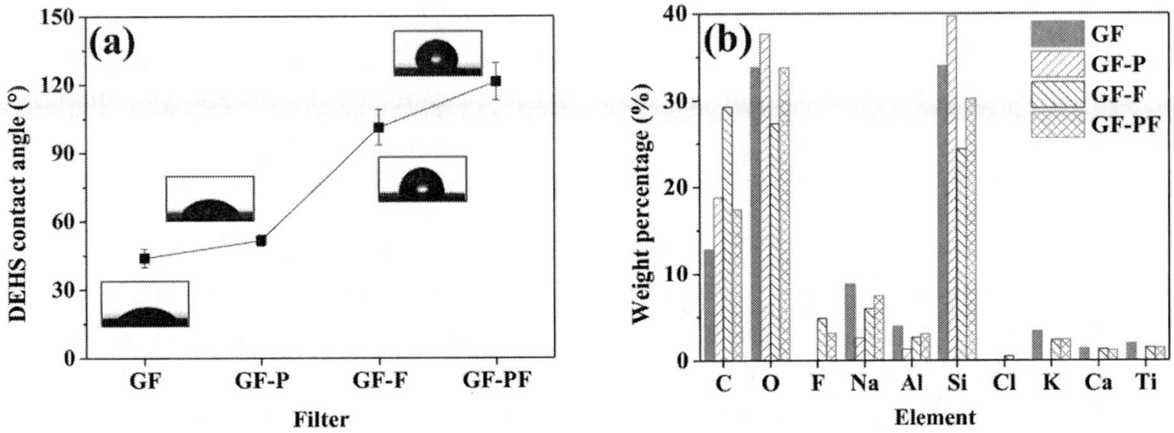

Fig. 3. Properties of different filter materials: (a) DEHS contact angle; (b) EDX.

3.2. Analysis of filtration performance

Fig. 4 shows the relationship between initial pressure drop and flow rate for filter GF-PF. As shown, initial pressure drop linearly increased with the air flow rate. Other filters illustrated a similar profile as this, suggesting that the fibrous materials were stably and evenly treated in some extents. By comparing the initial pressure drop under the experimental condition with face velocity remained at 0.12 m/s, as shown in Fig. 4, it was reasonable to conclude that there was little effect of surface modification method on the initial pressure drop of filter. This result was in good agreement with the previously mentioned result of similar pore size among different filters.

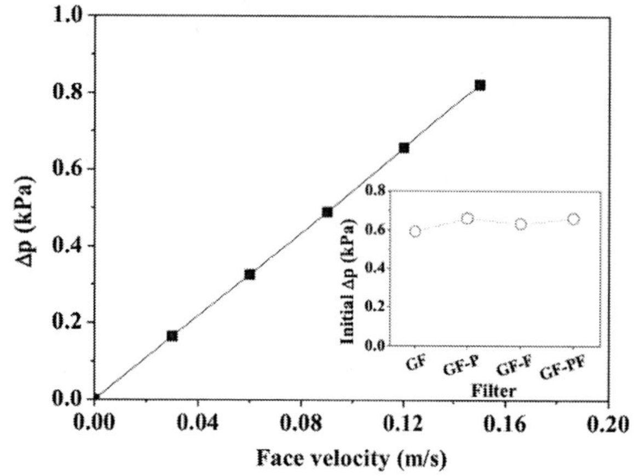

Fig. 4. Initial pressure drop of the filters.

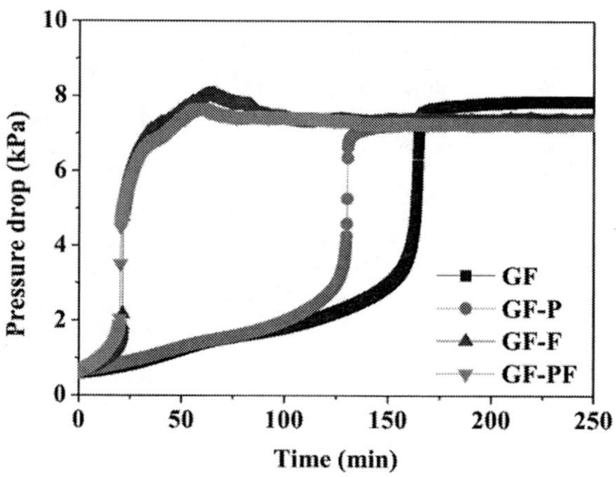

Fig. 5. Pressure drop evolution of the filters.

Fig. 5 shows the pressure drop evolution of the filters. Obviously, the steady-state pressure drop of three modified filters were almost the same, slightly lower than that of control filter. Because filter GF-F and GF-PF were both oleophobic, their pressure drop experienced a rapid increase after about 25 min from the experimental beginning, and then reached a steady state at approximately 100 min. By comparison, filter GF and GF-P presented a typical oleophilic trend, and the pressure drop increased slowly at early stage due to liquid channel transport, after which it appeared an earlier pressure drop jump for filter GF-P. Hence, it illustrates that a modification from oleophilic to oleophobic would fasten the evolution of pressure drop with time, which was more significant when the filter was relatively more oleophobic, while the steady-state pressure drop would basically not change.

Fig. 6 shows the steady-state filtration efficiency of the filters. As shown, filter GF-F and GF-PF had a significantly higher filtration efficiency for droplets in both particle size ranges when compared to the control filter. For filter GF-P, however, the efficiency was only slightly increased for droplets measured by the APS while decreased for those measured by the SMPS. It should be noted that filter GF-PF had the highest filtration efficiency for droplets measured by both particle sizers, suggesting that a combined modification of nanoparticles and fluoropolymer would maximize the filtration efficiency.

The steady-state quality factor was obtained after substituting filtration efficiency and pressure drop into Eq. (1), and the results are shown in Fig. 7. The quality factor of different filters illustrated a same fashion as the filtration efficiency shown in Fig. 6 due to their close pressure drop at steady state. Interestingly, filter GF-PF had the highest steady-state quality factor for droplets in both size ranges, meaning an overall best filtration performance. Therefore, it can be concluded that fiber surface modification by combination of nanoparticles and fluoropolymer, rather than only one of them, would most efficiently promote the filtration performance of the filter.

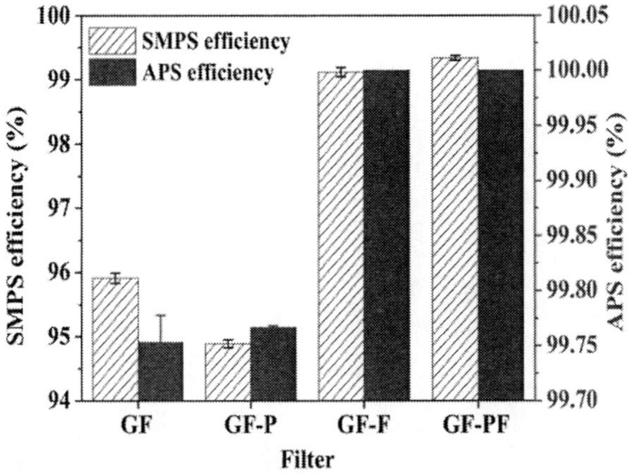

Fig. 6. Steady-state filtration efficiency of the filters.

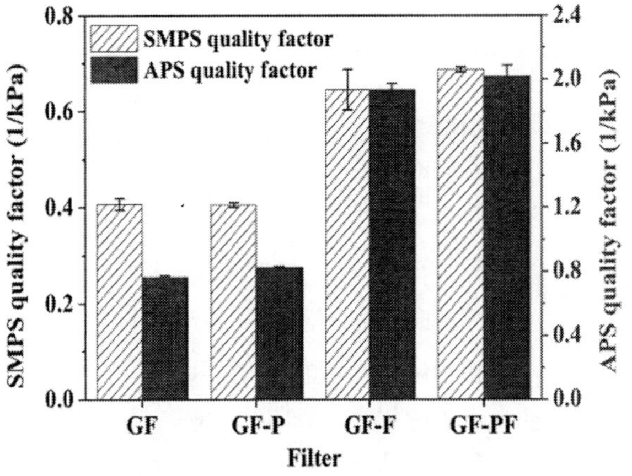

Fig. 7. Steady-state quality factor of the filters.

Fig. 8 shows the equilibrium saturation of the filters. For filter GF and GF-P, the saturation profile exhibited a typical oleophilic feature, where the second layer had an obviously higher saturation than the first layer. While for filter GF-F and GF-PF, the saturation of the second layer was largely lower than that of the first layer, on which a liquid film formed [5]. Note that the entire saturation of the last two filters was significantly lower than that of the former two filters. As a result, it is reasonable to suggest that the improvement of steady-state filtration efficiency of the filters was strongly related to the decrease of saturation, and the fiber surface modification was an efficient method to greatly reduce the saturation or liquid hold-up in the filter so that the filtration performance was promoted.

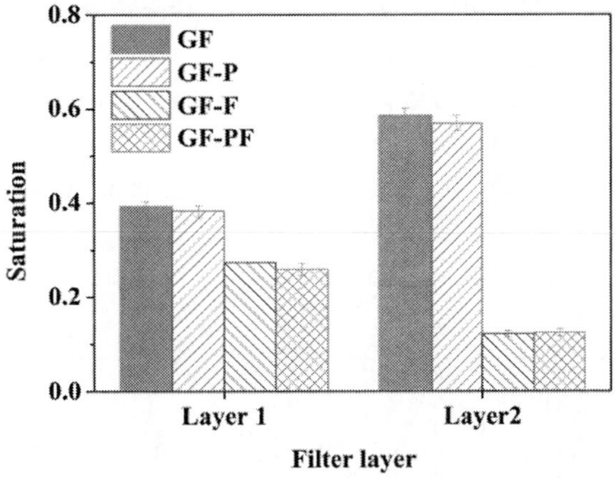

Fig. 8. Comparison of the saturation of different filters at steady state.

4. Conclusions

In this work the effect of surface modification on filtration performance of coalescing filters was experimentally conducted. The results showed that all surface modification methods used had little influence on the pore size and permeability of the filter. The oil wettability would not obviously change if the filter was only modified by nanoparticles, neither transferred from oleophilic to oleophobic. While modification by fluorination would effectively reduce the surface energy of the filter, thus increase the filtration performance. A combination of treatment by nanoparticles and fluoropolymer would produce a filter with highest oleophobicity, which was also of great advantage to the dispersibility and stability of nanoparticles on filter fibers. Moreover, the filter modified by this combined method exhibited an overall best filtration performance, due to significantly reduced liquid hold-up in the filter.

Acknowledgment

Authors acknowledge the financial support from National Natural Science Foundation of China under grant No. 51376196.

References

1. R. Mead-Hunter, A.J.C. King, B.J. Mullins, Aerosol-mist coalescing filters – A review, Sep. Purif. Technol. 133 (2014) 484-506.
2. G.M. Manzo, Y.R. Wu, G.G. Chase, A. Goux, Comparison of nonwoven glass and stainless steel microfiber media in aerosol coalescence filtration, Sep. Purif. Technol. 162 (2016) 14-19.
3. B.J. Mullins, R.D. Braddock, I.E. Agranovski, R.A. Cropp, R.A. O'Leary, Observation and modelling of clamshell droplets on vertical fibres subjected to gravitational and drag forces, J. Colloid Interface Sci. 284 (2005) 245-254.
4. R. Mead-Hunter, T. Bergen, T. Becker, R.A. O'Leary, G. Kasper, B.J. Mullins, Sliding/rolling phobic droplets along a fiber: measurement of interfacial forces, Langmuir 28 (2012) 3483-3488.
5. D. Kampa, S. Wurster, J. Buzengeiger, J. Meyer, G. Kasper, Pressure drop and liquid transport through coalescence filter media used for oil mist filtration, Int. J. Multiph. Flow 58 (2014) 313-324.
6. D. Kampa, S. Wurster, J. Meyer, G. Kasper, Validation of a new phenomenological " jump-and-channel " model for the wet pressure drop of oil mist filters, Chem. Eng. Sci. 122 (2015) 150-160.
7. C. Chang, Z.L. Ji, J.L. Liu, The effect of a drainage layer on saturation and liquid distribution of oleophobic coalescence filters, Sep. Purif. Technol. 194 (2018) 355-361.
8. C. Chang, Z.L. Ji, J.L. Liu, Pressure drop and saturation of nonwettable coalescing filters at different loading rates, AIChE J. 64 (2018) 180-185.
9. I.E. Agranovski, T. Myojo, R.D. Braddock, D. Jarvis, Combined wettable/non-wettable filter for mist purification, Chem. Eng. Technol. 24 (2001) 287-292.
10. B.J. Mullins, R. Mead-Hunter, R.N. Pitta, G. Kasper, W. Heikamp, Comparative performance of philic and phobic oil-mist filters, AIChE J. 60 (2014) 2976-2984.
11. G.M. Letts, P.C. Raynor, R.L. Schumann, Selecting fiber materials to improve mist filters, J. Aerosol Sci. 34 (2003) 1481-1492.
12. X. Wei, F. Chen, H.X. Wang, H. Zhou, Z.L. Ji, T. Lin, Efficient removal of aerosol oil-mists using superoleophobic filters, J. Mater. Chem. A 6 (2018) 871-877.
13. R.C. Brown, Air Filtration: An Integrated Approach to the Theory and Applications of Fibrous Filters, Pergamon Press, New York, 1993.

Particle Measurement in High-Temperature Gas based on Light Scattering

Li-feng LU[1,3], Xiao-lin WU [1,3], Zhong-li JI [1,2], Ming-xing WANG[1,3]

[1] Department of Chemical Engineering, China University of Petroleum, Beijing, P. R. China

[2] Department of Mechanical and Transportation Engineering, China University of Petroleum, Beijing, P. R. China

[3] Beijing Key Laboratory of Process Fluid Filtration and Separation, China University of Petroleum, Beijing, P. R. China

Abstract: In order to achieve the particle measurement at 650 °C, a single particle measurement method and device suitable for high-temperature gas were developed. The influence of gas temperature on optical sensor of optical particle counter (OPC) was investigated in detail based on geometric optics theory and heat transfer theory. The thermal expansion of aerosol tube along the airflow direction was caused by the heat conduction between the tube and the gas, resulting in the position of the light apertures to shift to obscure the partial light beam. The invar material can be adopted to eliminate the above mentioned disadvantages. According to the principle of gas state equation, the optical parameters (refractive index) of the gas at different temperatures change with the change of its state parameters. Based on the geometrical optics theory and the geometric mathematical relationships, the theoretical and experimental dynamic models of the optical measuring volume with the change of the gas temperature were established. The established models were then used to analyze the influence of gas temperature on the optical measuring volume. The results showed that the influence of the temperature on the optical system is negligible. Meanwhile, particle detection experiments were conducted at different temperature to achieve a good performance of OPC.

Keywords: High-temperature gas; Optical measuring volume; Dynamic model

1 Introduction

With the continuous improvement of environmental protection and the safety requirements for industrial production, the technology of high-temperature purification is drawing more and more attention, especially in the fields of clean coal power generation, coal gasification, petrochemical industry and metallurgy [1]. For example, nuclear power plants often produce a large number of low radioactive solid wastes, and the combustible wastes usually account for a considerable proportion. The incineration and pyrolysis are effective treatment techniques for the combustible waste, during which the temperature of flue gas is 600~800 °C. With the aim to purify the flue gas, the German Karlsruhe Nuclear Research Institute, the French nuclear power station and the Japanese nuclear power station have installed the silicon carbide ceramic filter tube, aluminum silicate fiber and alumina ceramic filter tube at gas temperatures above 650 °C, respectively, and investigated their properties properties [2-4]. These filters have a direct influence on the purification quality of gas and the particle concentration in the downstream. In an industrial cracker system, the particle concentration is required to reach 0.014 mg/m^3 after the filters; the Integrated Gasification Combined Cycle (IGCC) requires the particle size to be smaller than 1 um with a particle concentration of less than 2.2 mg/m^3 and in the range of 1~10 um with less than 5.8 mg/m^3, and to remove the dust size of larger than 10 um [5]. In a Shell Coal Gasification Process (SCGP), the particle concentration is required to be less than 20 mg/m^3 [6]. In order to detect the concentration and size of the particles in pipes in real time for calculating the separation efficiency of evaluating the purification performance of the filtration system, the particle detection technology is needed to be improved [7].

The widely-used particle detection devices in industry are mainly based on the principle of light scattering [8-10], and the particle size and concentration are obtained through an inversion of the scattering light intensity and particle size, which are suitable for the conditions of below 200 °C. The method of first cooling and then detecting can be adopted for the particle detection under higher-temperature conditions, but droplet condensation and sublimation will inevitably occur and result in an incorrect particle samples.

With the aim to achieve the particle detection at 650 °C, an optimization was conducted on the existing particle detector (OPC) in this paper. And based on geometric optics theory [11] and heat transfer theory, a detailed research was performed on the influence of gas temperature on OPC optical sensor, including the dynamic change of the aerosol tube and the optical measuring volume (OMV). Finally, the experiments were conducted on the on-line particles measurement of different temperatures.

2 Experimental setup and method

Figure 1 gives the experimental setup for detecting particles in high-temperature gas, mainly including the heating unit, optical detection unit and gas purification and processing unit.

The heating unit consists of the vibration particle feeder and the air heater. The gas brings the particle out of the feeder, and is heated up to 1000 °C by the heater, which is monitored quickly by a PID temperature control instrument with easy operation.

The optical detection unit is used to achieve the experimental measurement of the characteristics of the OPC optical system and the on-line detection of particles, including the optical sensor and the OPC control host. The optical sensor is shown in Figure 2, which is composed of an aerosol tube and optical system. Aiming to avoid the influence of high-temperature gas on optical system, the aerosol tube (shown in Figure 3) has been developed, and is in a distance of 15 mm to the optical lens. In addition, the insulation coating with lower thermal conductivity (-80-1700 °C) is adopted to treat the outer wall of the aerosol tube. A K-type micro thermocouple was bent and installed into the light aperture to monitor the gas temperature at the center of the aerosol tube in real time without affecting the airflow and optical paths. Figure 4 shows the optical system of the sensor, which is composed of an incident light path and an observed scattering light path. Both the paths are arranged in 90° with the same optical lens and elements [12-14], and the intersected area is the center of the aerosol tube to form an OMV [15]. The intersected area is enough small and allows particles to pass through it one by one [16, 17]. The OPC host consists of the light source system (wavelength: 300-700 nm) [18], the photoelectric detection system and the signal analysis and processing system. In the particle detection, particle is illuminated by the incident light when passing through the OMV, and the scattering light is collected and transmitted to the photoelectric detector to achieve the conversion of a pulse voltage signal from an optical signal. Finally, the tested particle size is calculated based on the calibration curve of particle size and voltage amplitude [14, 19].

The gas purification and processing unit is composed of a vacuum pump, a flow controller, a dust collector and a heat exchanger. The high-temperature gas form the aerosol tube is completely cooled during the heat exchanger, the flow rate of which is adjusted by the flow controller. In order to avoid the damage of particles on the pump, the dust is captured and collected by a filter in the dust collector.

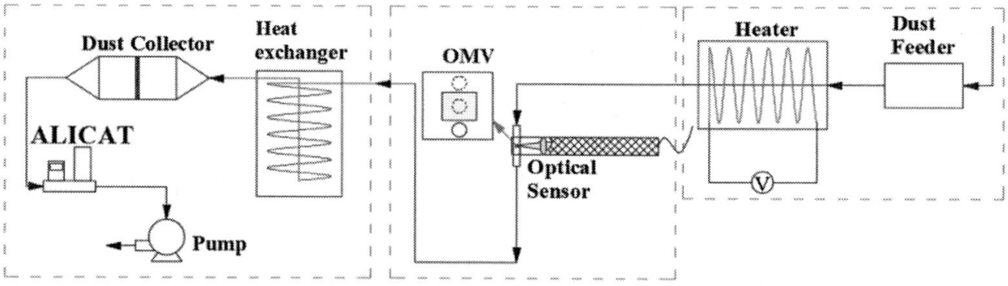

Figure 1. Experimental setup for detecting particles in high-temperature gas

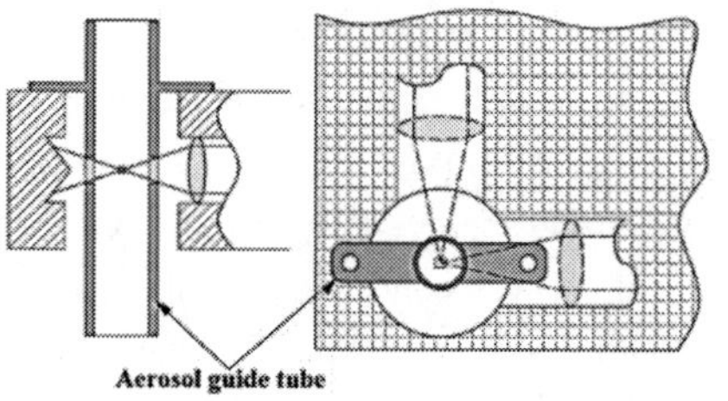

Figure 2. Optical sensor

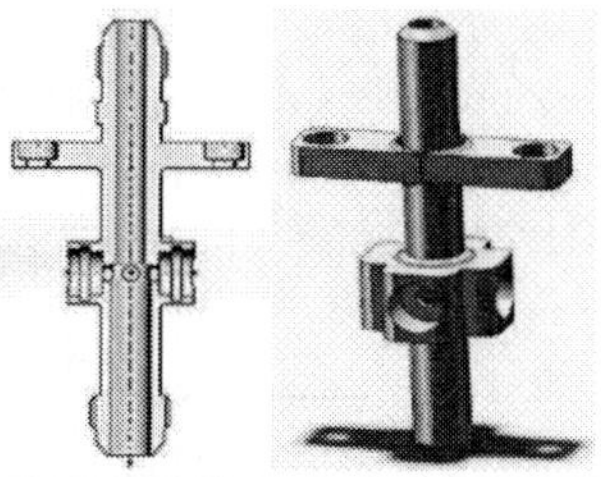

Figure 3. Aerosol tube for high-temperature gas

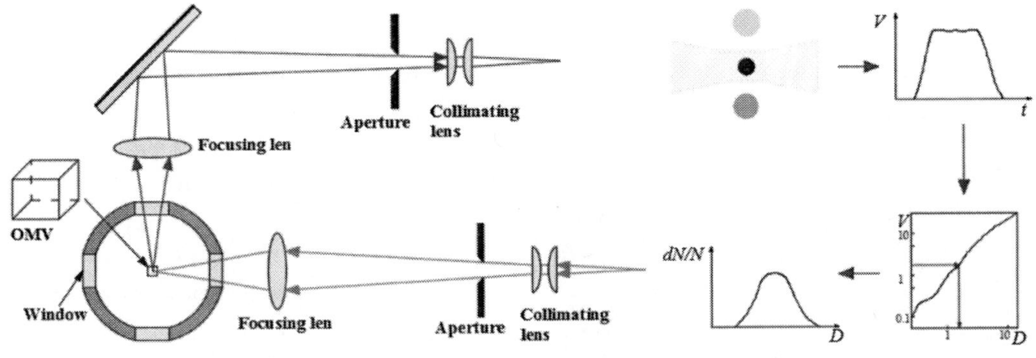

Figure 4. Optical system and principle of OPC

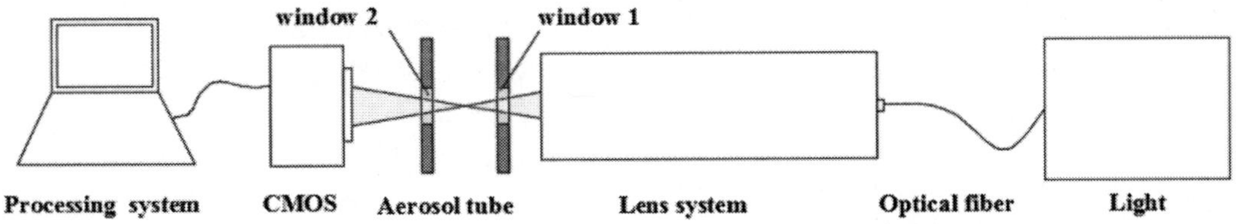

Figure 5. Evaluation and analysis device for light beam

3 Analysis and discussion

3.1 Influence of temperature on aerosol tube

The aerosol tube shown in Figure 3 has four apertures along the center axis, in which the glasses with a thickness of 2 mm were installed, as shown in Figure 4. When the gas flows in the tube, it can reduce the damage of high-temperature gas to the optical system elements. But the heat resistance of the tube has a direct influence on the imaging of the optical system at the center of the aerosol tube.

Figure 5 gives an evaluation and analysis device for light beam, including the light source system, the light path system, the aerosol tube and the light beam analyzer (CINOGY CMOS-1201, Germany), which can be equipped at the optical testing unit as shown Figure 1. For the gases at different temperatures, the light passes through the glasses in the aerosol tube and images on the CCD of CMOS-1201 to detect the spot of the light beam in real time.

After eliminating the background noise, the spot image at different temperatures can be acquired. When the aerosol tube is made of 304 stainless steel, the measurement results are shown in Figure 6. It can be found that the shape of the light spots is distorted and shows the imperfection of the imaging with an increase in gas temperature. That's due to thermal expansion along the airflow direction caused by the change of gas temperature, which results in the shift of light apertures position to obscure the partial light beam. A bad result must be acquired for smaller particle size and lower particle count efficiency because of the decreased incident light. Meanwhile, the shape of the OMV was changed to greatly reduce the performance of OPC detection by a failure of eliminating the error of edge detection in optical system.

With the aim to eliminate the above mentioned disadvantage, the invar material with a linear expansion coefficient of 1.6×10^{-6} m/°C was adopted for high-temperature experiment. The results are shown in Figure 7. At different temperatures, no deformation was found in the shape of light spots, indicating that the material can meet the testing requirement for high temperature up to 650 °C with a good heat resistance.

Table 1. Comparison of the application conditions of the two materials

Material	Light divergence angle /°	Gas temperature /°C
304 (Stainless steel)	<10	<200
4J36 (Invar)	<30	<700

The application conditions of the two materials are

summarized in Table 1. Compared with the stainless steel, the invar material is of great significance for the development of high-temperature detection device.

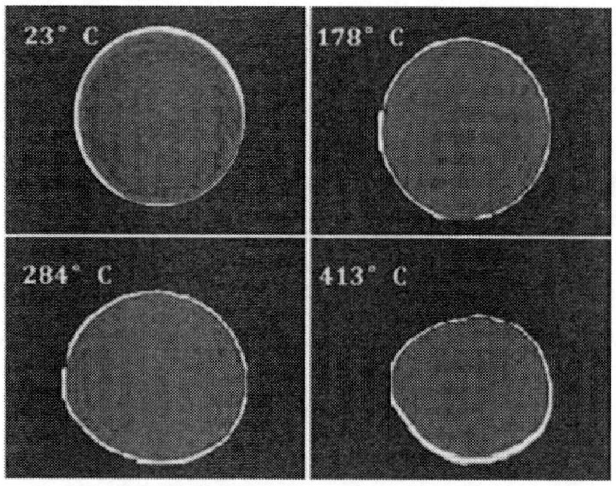

Figure 6. The spot images measured at four different temperatures of gas for 304 stainless steel.

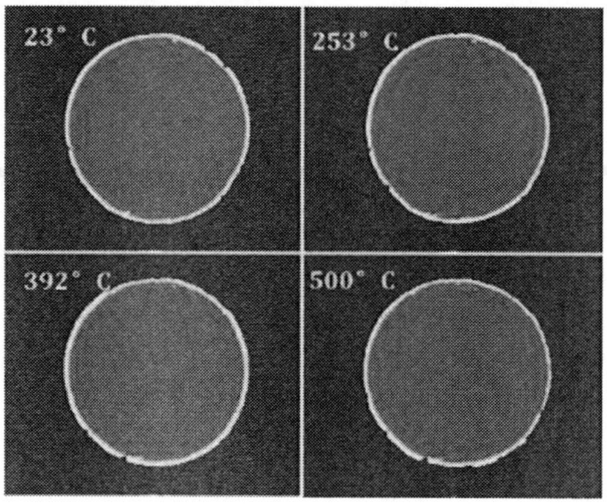

Figure 7. The spot images measured at four different temperature of gas for 4J36 (Invar).

3.2 Influence of temperature on optical system

In general the optical sensing system in the particle detection device is only applied at atmospheric temperature, but can also satisfy the temperature up to 150 °C. Under high-temperature condition, the gas volume is going to expand with the reduction in the gas density. The transmission characteristic of light beam could be changed, resulting from the decrease in the optical performance of the sensor. In order to investigate the influence of temperature on optical system, the specific causes are analyzed by theoretical method, and the extent of its influence is determined through experimental verification.

3.2.1 Theoretical method

By calculating the refractive index of the gas at different temperatures, the change of the OMV with the temperature can be obtained.

a) Calculation of gas parameters

The gas medium is air, so its refractive index can be calculated by the formula proposed by Rueger at the working temperature, given by:

$$n_{ph} = \frac{273.15}{1013.25} \cdot P \cdot N_{sph} \cdot (a_0 + \frac{a_1}{T} + \frac{a_2}{T^2} + \frac{a_3}{T^3} + \frac{a_4}{T^4}) \times 10^{-6} + 1 \quad (1)$$

Where P is the gas pressure; a_0, a_1, a_2, a_3 and a_4 are the correlation coefficient and constant; N_{sph} is the gas refractive index at atmospheric temperature, given by:

$$N_{sph} = (n_{sph} - 1) \times 10^6 = 287.6155 + \frac{1.62887}{\lambda^2} + \frac{0.01360}{\lambda^4} \quad (2)$$

The wavelength range is 400-700 nm. The gas refractive indexes at different temperatures in the range of 25-650 °C are calculated by Eq. 1 and 2. The results are shown in Figure 8. It can be found that the refractive index of air decreases with the increase in the gas temperature. And it can be inferred that the light divergence angle at high temperature is going to increase compared with that at atmospheric temperature. Consequently, the position of the beam focus is offset, leading to larger optical measuring volume, as shown in Figure 9.

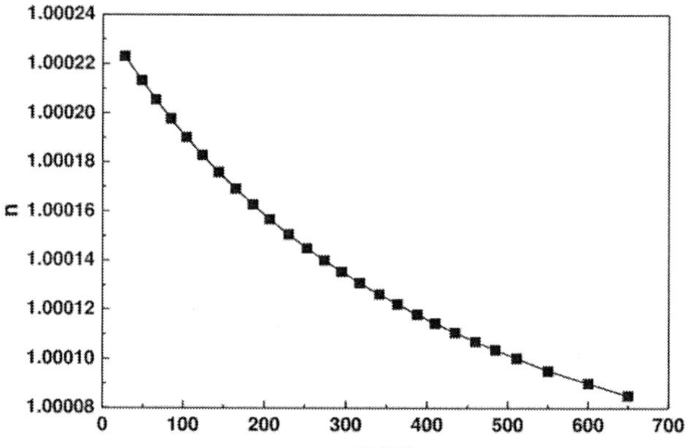

Figure 8. Gas refractive index at different temperature

In order to analyze the above unfavorable influence, a model of the OMV and gas temperature is needed to be built, mainly including the functions of the determination of the optical parameters and the calculation of the dynamic variable.

b) Calculation of optical parameters

On the basis of the Lorentz-Lorenz formula and actual gas state equation, the relationship between gas pressure and refractive index is given by:

$$\frac{n^2-1}{n_0^2-1}=\frac{\rho}{\rho_0}=\frac{PT_0Z_0}{P_0TZ} \tag{3}$$

Where T_0, P_0, ρ_0, Z_0 and n_0 are the gas temperature, pressure, density, compression factor and refractive index under atmospheric condition, respectively; T, P, ρ, Z and n are the gas temperature, pressure, density, compression factor and refractive index under high-temperature condition, respectively.

The light divergence angle is calculated by the refraction law shown below:

$$n_0' \sin\alpha = n \sin\beta \tag{4}$$

Where n_0' and α are the air refractive index and light divergence angle under atmospheric condition; β is the light divergence angle under high-temperature condition.

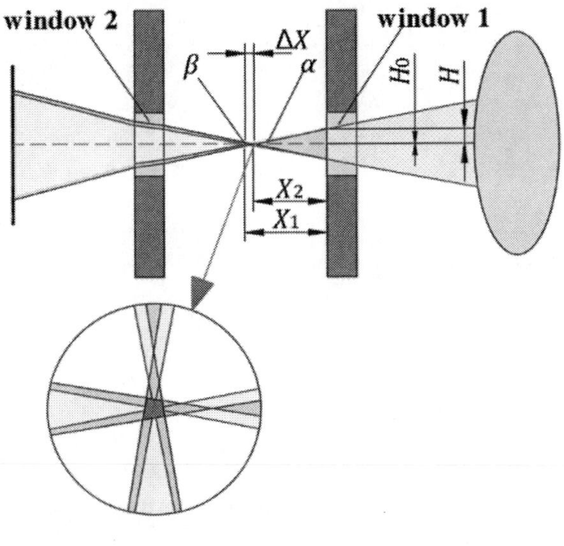

Figure 9. Schematic of the enlargement of the OMV

c) Calculation of dynamic variable

Based on the geometric optical theory and the mathematic relationship, the distance between the light focus position and the glass window 1 is triangular with the equivalent spot radius of the inner side of the glass window 1. According to X_1 and X_2 (the distance between the light focus position and the inner side of the glass window 1 under atmospheric condition and high-temperature condition), the theoretical model of the OMV with gas temperature was built. X_1 and X_2 are calculated by:

$$X1=\frac{H-H_0}{\tan\alpha} \tag{5}$$

$$X2=\frac{H-H_0}{\tan\beta} \tag{6}$$

The change of the light focus position (ΔX=$X1$-$X2$) can be calculated by Eq. 5 and Eq. 6.

The change of the equivalent spot diameter of the OMV can be calculated based on the triangle relation, as shown below:

$$\Delta D = 2\Delta X \tan\beta \tag{7}$$

Eq. 6 and Eq. 7 are the formulas of the theoretical dynamic model of the OMV with the change of the gas pressure.

Using the model, the optical parameters at different gas temperatures in the range of 25 ~ 700°C were calculated with three light divergence angles (4.02°, 9° and 13°, tested under atmospheric condition, respectively). The results are presented in Figure 10. It can be found in Figure 10 that ΔX increases with the gas temperature with approximate two patterns. In addition, ΔD changes by a smaller margin (less than 1 um). However, ΔD and ΔX satisfy a trigonometric relation. It can be also noted that the enlargement rate of the OMV volume (δ) has a small increase with the increase in gas temperature. Under the constant temperature, δ is going to increase for the larger light divergence angle, but it is still relatively smaller (<1%) compared with the OMV at atmospheric condition. Consequently, the influence of the gas temperature on the OMV can be ignored based on the above theoretical analysis.

3.2.2 Experimental method

In order to verify the reliability of theoretical analysis, an experimental dynamic model of the OMV with gas temperature variation was established based on the image analysis and the mathematical geometry relations, including the measurement of the change of light spot, the calculation of the geometric relation and the calculation of the relative variable.

By the refraction law, the changes of light spot on the inner and outer side of the glass window 2 under high-temperature condition (ΔD and $\Delta D'$) meet ΔD=$\Delta D'$. Taking the light focus position as the central axis, a triangle relation was formed by the distances between the light -

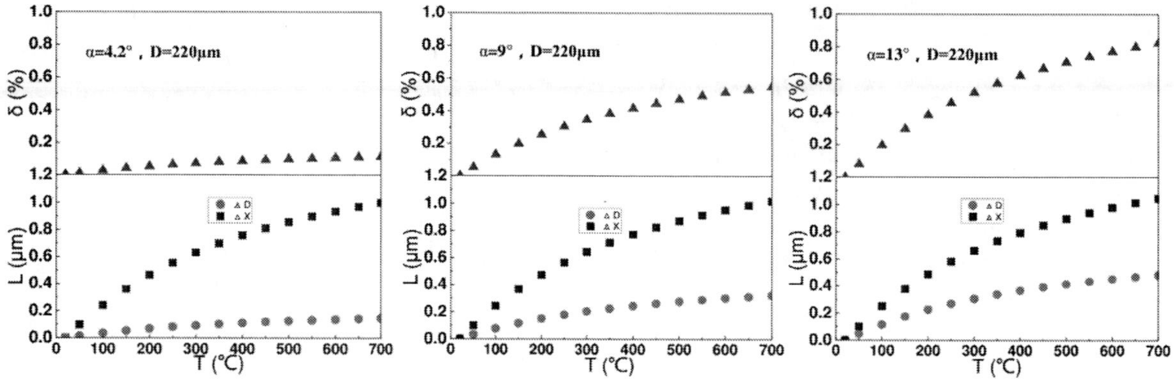

Figure 10. Results calculated based on theoretical dynamic mode with three light divergence angles (4.02°, 9° and 13°)

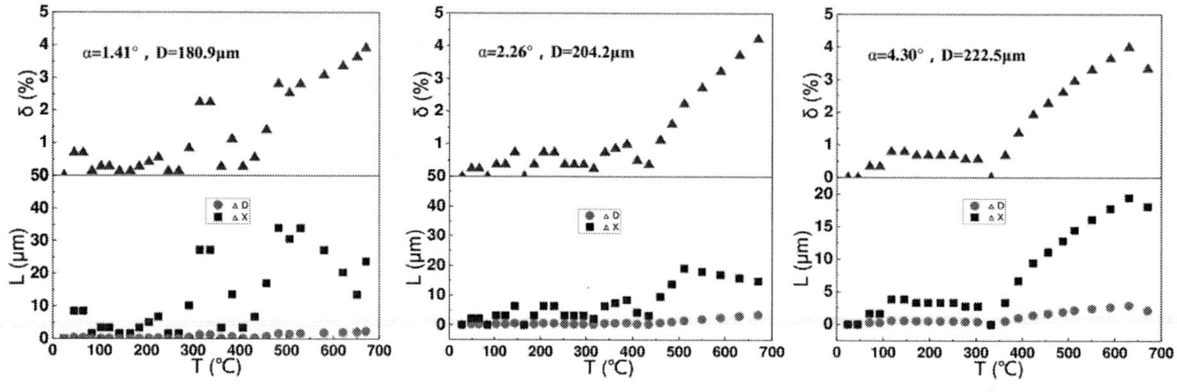

Figure 11. Results calculated based on experimental dynamic model

focus position and the inner side of the glass windows 1 and 2 (x and y), the difference of the light spot equivalent radius on the inner side of the glass windows 1 and 2 (H and h, $h = H - \frac{\Delta D}{2}$) and the equivalent radius of the OMV (H_0). Therefore, we have the following equations.

$$\frac{H - H_0}{h - H_0} = \frac{y}{x} \tag{8}$$

$$x + y = k \tag{9}$$

$$\beta = \arctan\left(\frac{h - H_0}{x}\right) = \arctan\left(\frac{H - H_0}{y}\right) \tag{10}$$

where β is the light divergence angle under high-temperature condition; k is the distance of the inner sides of the glass windows 1 and 2.

The changes of ΔX and ΔD under high-temperature condition to those under atmospheric condition are shown as follows:

$$\Delta X = \frac{y - x}{2}, \text{ or } \Delta X = y - \frac{k}{2}, \text{ or } \Delta X = \frac{k}{2} - x \tag{11}$$

$$\Delta D = 2\Delta X \tan\beta = \Delta D' \tag{12}$$

Eq. 11 to Eq. 12 are the formulas of the experimental dynamic model.

Using the above established experimental dynamic model, the ΔX, ΔD and δ of four cases in Table 2 were calculated, and the results are shown in Figure 11. It can be found that these three parameters increase obviously with the gas temperature increasing, but cannot meet the two function growth mode as the results of theoretical model. That may be caused by the reasons listed below: 1) the fluctuation of gas flow in the pipeline causes the unstable gas temperature field; 2) the beam quality analyzer is slightly influenced by other vibration sources; 3) the aerosol tube is affected by the gas flow fluctuation. But according to the experimental results, the maximum values of ΔX and ΔD are very small (less than 50 μm and 5 μm at the temperature below 700 °C, respectively) compared with the equivalent diameter of the OMV. Consequently, the ratio of the volume of the OMV is less than 5%. To sum up, the gas temperature has greatly weakened influence on the OMV, which can be ignored in actual technical development.

Table 2. Optical parameters of experimental model

number	light divergence angle /°	spot equivalent diameter /μm
1	1.41	180.9
3	1.94	285.1
4	2.26	204.2
5	4.30	222.5

3.2.3 Experiments

The performance of the optical sensor is investigated at different temperatures up to 650°C. The experimental results showed that the temperature of the optical sensor is around 50°C with the gas temperature in the range of 200-300 °C, and is going to increase about 10 °C at the gas temperature of 400-500 °C. The temperature is in the controllable range with a maximum value of 70 °C under the gas temperature of 650 °C. Therefore, the developed optical sensor can be used for the detection of particles in high-temperature conditions for a long time.

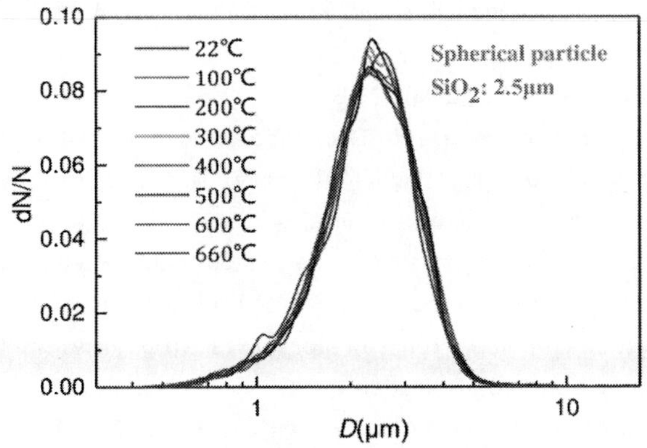

Figure 12. The results obtained under high-temperature condition

In the experiment of particle detection, the standard silica spherical particles with a strong temperature resistance ($D_{particle}$=2.5 μm) were tested at different gas temperatures. The testing time is 60 s. Figure 12 shows the test results. It can be found that all of the particle size distributions are in good agreement and have a good normal distribution, which meets the test results of single particle size, and the peak particle size is about 2.5 μm. Consequently, the optical sensor worked well for the particle detection under higher-temperature condition.

4 Conclusion

The emission of high-temperature flue gas causes serious air pollution and partial waste of available heat. In order to meet the relevant emission standards, additional purification units (high-temperature ceramic filters, etc.) are installed in the industrial process to collect the dusts in the flue gas. As the purification units are important for controlling the dust concentration, it is necessary to calculate the separation efficiency through a real-time measurement of the dust concentration in the flue gas in the upstream and downstream of the purification equipment.

In present work, a single particle measurement method and device suitable for high-temperature condition were developed. In order to complete the particles measurement at 650 °C, the beam images in optical sensing system under different temperature conditions were measured in real time by CMOS sensor. It was found that the beam images were deformed when the light divergence angle increased to more than 10°, and the deformation became more serious as the temperature increased. Depending on the principle of heat transfer, thermal expansion along the airflow direction caused by heat conduction between the duct and the gas is the main cause of this phenomenon. By changing the aerosol tube material t o one with an ultra-small linear expansion coefficient (Invar: 4J36), the optical system can be applied to the condition with temperature up to 700 °C, and in particular the light divergence angle extends to 30°.

According to the principle of gas state equation, the optical parameters (refractive index) of the gas at different temperatures change with the change in its state parameters. According to the results calculated by the formula of Ruegur's atmospheric refractive index, it was found that the refractive index of the gas decreases with the increase in temperature, which leads to the increase in the divergence angle of the beam in the aerosol tube, and shift of the position of the optical beam. Based on the geometrical optics theory and the geometric mathematical relationships, the theoretical dynamic model of the optical measuring volume with the change of the gas temperature was established. By calculating the amount of the beam focus offset and the volume of the optical measuring volume at different temperatures, it was found that the optical measuring volume increases with the temperature increasing. However, when the gas temperature was 650 °C, the increase rate of optical measuring volume was extremely small, so the influence of the temperature on the optical system can be negligible.

The performance of the whole detection system was tested by using 2.5 μm standard spherical silica particles. The satisfying measuring results were obtained and they could be applied to the actual on-site measurement at the temperature of 650 °C.

Acknowledgement

The authors thank the supports of "The National Key Research and Development Program of China (NO.2016YFB0601100)"; and would also like to thank the laboratory and staff of the Beijing Key Laboratory of Process Fluid Filtration and Separation, China University of Petroleum.

Reference

[1] J. Meyer, H. Umhauer, K. Reuter, A. Schiel, G. Kasper, M. Förster, Concentration and size measurements of fly ash particles from the clean gas side of a pressurised pulverised coal combustion test facility, Powder Technology, 180 (2008) 57-63.

[2] H. Leibold, F. Dirks, V. Rüdinger, Particulate emissions from a LLW incinerator and off-gas cleaning with a new type of ceramic candle filter, Waste Management, 9 (1989) 87-94.

[3] B.L. Perkins, Incineration facilities for treatment of radioactive wastes: a review, Incinerators, (1976).

[4] D. Fino, N. Russo, A. G. Saracco, V. Specchia, Multifunctional Filter for Treatment of the Flue Gases from Municipal Waste Incinerators, Industrial & Engineering Chemistry Research, 44 (2005) págs. 9542-9548.

[5] Scheibner, Wolters, Schumacher hot gas filter long-term operating experience in the nuon power buggenum IGCC power plant, (2002).

[6] J.M. Wang, Shell coal gasification technology and its application in China, Coal Quality Technology, (2006).

[7] Q. Xu, Z. Ji, L. Yang, Performance assessment of high temperature flue gas on-line particle size analyzer, Ciesc Journal, 63 (2012) 3506-3512.

[8] B.Y.H. LIU, V.A. MARPLE, K.T. WHITBY, N.J. BARSIC, Size distribution measurement of airborne coal dust by optical particle counters, Aihaj, 35 (1974) 443-451.

[9] R. Maus, F. Löffler, Size analysis of airborne bacteria by an optical particle counter and an impactor, Journal of Aerosol Science, 25 (1994) 141-142.

[10] B. Sachweh, H. Büttner, F. Ebert, Improvement of the resolution and the counting accuracy of an optical particle counter by fast digital signal recording, Journal of Aerosol Science, 20 (1989) 1541-1544.

[11] E.W. Max Born, Principles of Optics, Cambridge University Press, Cambridge, 1999.

[12] S. Hochrainer, Further developments of the particle counter sizer PCS-2000, Journal of Aerosol Science, 31 (2000) 771-772.

[13] B.S. Li, Z.L. Ji, L. Mei, An approach for correcting particle size distributions measured by the optical particle counter welas 2000, Aip Conference, 2010, pp. 686-689.

[14] J. Stier, M. Quinten, Simple refractive index correction for the optical particle counter PCS 2000 by Palas, Journal of Aerosol Science, 29 (1998) 223-225.

[15] H. Umhauer, Particle Size Distribution Analysis by Scattered Light Measurement Using an Optically Defined Measuring Volume, Journal of Aerosol Science, 14 (1983) 765-770.

[16] B. Sachweh, H. Umhauer, F. Ebert, H. Buttner, R. Friehmelt, In situ optical particle counter with improved coincidence error correction for number concentrations up to 107 particles cm-3 quantification and elimination of the effect of particle shape and structure, Journal of Aerosol Science, (1998) 1075-1086.

[17] D.I.J. Raasch, D.I.H. Umhauer, Errors in the Determination of Particle Size Distributions Caused by coincidences in optical particle counters, Particle & Particle Systems Characterization, 1 (1984) 53-58.

[18] H. Buettner, Measurement of the Size of Fine Nonspherical Particles with a Light-Scattering Particle Counter, Aerosol Science & Technology, 12 (1990) 413-421.

[19] D. Holve, S.A. Self, Optical particle sizing for in situ measurements Part 2, Applied Optics, 18 (1979) 1632-1645.

508245 An Experimental Investigation of Single Droplet Drying Above Boiling Point

Abstract:

An experimental method has been developed to enable dried particle structure, properties and drying rate to be predicted based on droplet drying history, particularly focusing on effects driven by boiling including bubble nucleation, growth and collapse. The method involves a single droplet drying rig which will be used to investigate drying kinetics above boiling point and across a broad range of drying conditions, drop size and composition space. The drying rig was improved as compared to previous apparatuses reported in literature, in that it is designed to measure droplet drying rates and morphological changes at high temperatures and under conditions that more closely resemble those found in industrial applications. The aim is to conduct drying experiment for droplets as small as 300μm diameter, in an air stream which temperature may be as high as 350°C.

Drying kinetics and morphological evolution will be determined, particularly focusing on the second drying stage as well as inflation/deflation periods. Initially, the constructed drying rig is checked and evaluated by conducting experiments on droplets of pure water. Afterwards, experiments will be conducted on solutions of different materials such as Hydroxypropyl Methylcellulose (HPMC), sodium sulphate and sucrose. An assessment of morphologies, transformations and drying behaviour under isothermal drying would be completed at a wide space of process condition, initial water content, initial drop size and composition space. The investigation will look to link material properties, particle size, initial moisture content and drying history to morphology. Results will be compared with single droplet drying models reported in literature.

An initial investigation has focused on mapping the morphological drying behaviour of HPMC, sodium sulphate and sucrose. It has shown that the three materials show different behaviour; sucrose showed deflation and inflation cycles, sodium sulphate droplets supersaturated and crystallized while HPMC showed inflation and rupture. Differences in drying behaviour is due to difference in material and transport properties. Therefore, the next stage of this work is focused on conducting some physical properties measurement at different water concentrations and temperature. In addition, different materials will be dried on the rig and the drying behaviour will be determined. The experimental data from this work will be used to develop, a universal morphology regime map which will link material properties, particle size, initial moisture content and drying history to morphology.

508482 Characterising Powder Flow in Dynamic Processes

Abstract:

Flow inconsistency of cohesive powders is a major issue for process industries. It is often difficult to predict and control flow of cohesive powders to enable efficient and reliable process operations particularly for dynamic processes such as mixing and filling. Many quantitative and qualitative flow measurement techniques exist that are used for measuring powder flowability under quasi-static conditions. However, there are few available techniques capable of assessing powder flowability under dynamic conditions. The FT4 Powder Rheometer (Freeman Technology), which drives a rotating blade along the vertical axis through a column of powder contained within a cylindrical vessel, is one of the few available devices that is capable of measuring powder flow in a dynamic state [1]. The flow resistance of the rotating blade as it penetrates the powder bed is characterised by flow energy, however, there is lack of understanding in how to relate this measurement to process design.

This work focuses on determining the relationship between the torque and shear stress in the FT4. The effect of particle size on the external flow profile around the blade has been investigated experimentally for glass beads using a high speed camera. The material investigated shows a decrease in downward specific flow energy with increasing particle size; indicating that the increased shear resistance in the bulk for smaller particle sizes exists under dynamic conditions. Particle Image Velocimetry (PIV) analysis indicates an increase in the volume of particles displaced by the blade with increasing particle size for both upward and downward blade movement. An increase in impeller tip speed leads to a reduction in flow energy for this material, and a reduction in the volume displaced by the blade. The prevailing stresses and strains as well as the volume of particles displaced by the blade are estimated by Discrete Element Method (DEM). DEM allows for the determination of a relationship between the flow energy and/or shear stress with particle properties and blade velocity and direction. Furthermore, Positron Emission Particle Tracking (PEPT) analysis of the FT4 has been carried out to determine the radial and axial displacement of particles with respect to the bed height, and to validate the DEM results.

508671 Impact of Non-Spherical Projectiles on Granular Media

Abstract:

Recently, several studies involving numerical and experimental methods have focused on the study of impact dynamics in both dry and wet granular media. An example of such instances can be an animalâ€™s foot impacting sand during its locomotion. Understanding the physics behind the interaction between granular media and animal feet can be helpful in optimizing locomotion parameters for bio-inspired robots. However, most of the studies involving impact dynamics considered spherical projectiles under different conditions, while characteristic models should involve more complex shapes. Impacting different geometries with conserved density, volume, and velocity on a granular bed may experience contrasting drag forces upon penetration. This is due to the difference in the surface areas coming in contact with the granular media. In this work, non-spherical geometries of following shapes were considered: cylinder, triangular prism, hexagonal prism and cuboid. These geometries were impacted on a loosely packed non-cohesive dry granular bed with the same impact velocities. The effect of surface area in contact with the granular media was analyzed for impact, with the help of penetration depth measurements. We also tested whether the power law for spherical projectiles is applicable for non-spherical projectile impacts.

508664 Computational and Experimental Shear Cell Study with Rigid Cylindrical Particles

Abstract:

Flow properties of particles are commonly assessed through shear cells. Both the Schulze ring shear tester (SRST) and Jenike shear tester (JST) measure the interparticle stresses which are important to the design of many granular flow processes. This work uses precision-cut rigid cylinders of various aspect ratios and sizes to compare the measured stresses between both testers. It is found that the shearing results for both testers are similar, which is important for future work involving flexible biomass materials.

Previous work has sought to use shear cell results to validate various aspects of discrete element method (DEM) simulations. The SRST is recreated in a DEM simulation with rigid glued-sphere cylindrical particles to assess the contact detection method by comparing the experimental and computation results. Also, a parameterization study is performed to examine the effects of coefficient of restitution, coefficient of friction, and other particle properties on the measured steady-state stress. Overall, this serves as an efficient validation of the contact detection method which will be used to simulate complex particle shapes.

508865 Defluidization Behaviour of Industrial Reactive Powders at High Temperature

Abstract:

Fluidized bed reactors are used in a wide range of industrial chemical processes, and they are particularly useful in high temperature systems thanks to their ability to provide high heat transfer rates and rapid solids mixing which lead to isothermal and more controllable conditions. Their applications spread the environmental, chemical, energy and process industries. For example, they are a key technology in the petroleum industry, catalytic chemicals synthesis processes, combustion and gasification of solid fuels (coal, wastes and biomass), metals productions and many more.

Use of fluidized beds in combustion-type operations may lead to particles sintering and bed agglomeration, which again may result in defluidization, unscheduled process downtime and additional costs. However, the details of the physical mechanisms behind such phenomena are still not entirely understood. Despite the countless researches that have been done so far, such processes still disclose unrevealed features and challenge the researchers worldwide.

The objective of this paper is to assess the effect of high temperature on the flow properties and fluidization behaviour of mixtures of carbon coke and two different titanium ores, namely synthetic and natural rutile. To this end, a simple approach for the early detection of agglomeration in the regimes of incipient fluidization -where the carbon combustion causes defluidization- will be followed. Such a monitoring approach is based on the simultaneous measurements of local temperatures and the pressure drop across the particles bed.

507739 Manufacturing and Characterization of Spherical Blend PBT-PC Particles for Additive Manufacturing

Abstract:

Polymer powder-based additive manufacturing technologies often yield excellent mechanical and functional part qualities. One of these technologies is selective laser sintering (SLS), a process employing micron-sized polymer particles, which are selectively fused by a laser [1]. While hardly any boundaries regarding part design exist, there are quite some restrictions concerning the variety of commercially available SLS materials. For the time being, the most widely used polymeric material for SLS is polyamide 12 (PA12) with a market share of roughly 95% [2]. Hardly any blend materials are available; mostly , the processing of multi-material systems via SLS is achieved by simply mixing different polymer powder materials [3].

In this contribution, a new approach to blend particles is introduced. Via co-grinding and subsequent rounding, manufacturing of spherical particles consisting of semi-crystalline polybutylene terephthalate (PBT) and amorphous polycarbonate (PC) thermoplastic is possible. The polymers are jointly comminuted and mixed in a planetary ball mill and the product particles are rounded in a heated downer reactor [4,5]. The thereby achieved intermixed, but not transesterified, PBT-PC composite particles are characterized regarding size and morphology via electron microscopy. Intermixing of the polymers is investigated via staining techniques and Raman spectroscopy. Thermal properties of the blend particles are analyzed via DSC, also in regard to transesterification in SLS process-like conditions. Furthermore, flowability and size distribution of the blend particles are examined. This way, we show the high potential of our process route [4] for the production of novel polymer blend particle systems for SLS.

References:

[1] I. Gibson, D. Rosen, B. Stucker, Additive manufacturing technologies, 2. ed., Springer, New York, NY, 2015.

[2] T.T. Wohlers, Wohlers Report 2016, Wohlers Associates, Fort Collins, Col., 2016.

[3] D. Drummer, K. Wudy, F. Kühnlein, M. Drexler, Polymer Blends for Selective Laser Sintering: Material and Process Requirements, Phys. Procedia. 39 (2012) 509–517. doi:10.1016/j.phpro.2012.10.067.

[4] J. Schmidt, M. Sachs, C. Blümel, B. Winzer, F. Toni, K.-E. Wirth, W. Peukert, A novel process route for the production of spherical LBM polymer powders with small size and good flowability, Powder Technol. 261 (2014) 78–86. doi:10.1016/j.powtec.2014.04.003.

[5] J. Schmidt, M. Sachs, S. Fanselow, M. Zhao, S. Romeis, D. Drummer, K.E. Wirth, W. Peukert, Optimized polybutylene terephthalate powders for selective laser beam melting, Chem. Eng. Sci. 156 (2016) 1–10. doi:10.1016/j.ces.2016.09.009.

506658 Creating Tuneable Agglomerates via 3D Printing

Abstract:

It is essential to offer a universally accepted way to systematically evaluate agglomerate breakage propensity and mechanisms. Discrete Element Method (DEM) simulation is a commonly used tool but is limited by the lack of identical, controlled agglomerates to test and validate simple models, let alone replicate the complex structure of real industrial agglomerates.

Our research presents a systematic approach to produce 3D printed agglomerates with tuneable properties. Agglomerates with different mechanical properties and structures were designed and printed by an Objet500 Connex 3D printer. The tensile and shear strength of inter-particle bonds was characterized using Instron tester. Breakage results of agglomerates with simple cubic structures and the design of agglomerates with irregular structures were presented. This 3D printing method will allow more rigorous testing of agglomerate breakage models. Discrete Element Method (DEM) has been used to simulate the agglomerate crushing using the Timoshenko Beam Bond Model (TBBM) with bond properties matching the 3D printed agglomerates. For both two agglomerate structures used in this research, the simulation and experiment showed similar breakage patterns. However, the simulation results significantly underestimated the compressive loads of spherical random structured agglomerates. In order to improve the simulations of compressive granule breakage, the influences of bond geometry definitions and non-linear behaviour of the bond contacts need to be explored.

508699 Effect of Particle Size Distribution and Wettability on Penetration Behavior, Granule Formation and Granule Properties in Single Drop Granulation

Abstract:

In wet granulation, single drop granulation (where one drop forms one granule) can simplify the complexity and reveal the fundamentals of the interaction between the liquid and the powder bed. Combining drop penetration time, granule formation mechanisms and the characterization of granule properties is a novel study to achieve a comprehensive map of the single drop granulation process, which is relevant in particulate applications like pharmaceuticals, detergents, agricultural chemicals and catalysts.

In this work, single drop impact of liquid on a static powder bed was studied to investigate the granule formation mechanism, droplet penetration time, and the characterization of granule properties (morphology, internal structure and content uniformity). Water was used as the liquid binder and two pharmaceutical powders, microcrystalline cellulose (MCC) (mean particle size ~54.8 Î¼m, contact angle with water ~36Â°) and acetaminophen (APAP) (mean particle size ~9.74 Î¼m, contact angle with water ~89°), were mixed to make heterogeneous powder beds with varying compositions. The complete drop impact and penetration was recorded with a high-speed camera. The granule morphology, its internal structure and content uniformity were characterized by the prism method with image analysis, micro-CT and UV-vis spectrometry, respectively. Besides the current MCC and APAP system, MCC and APAP with other particle sizes and relative contact angles will also be studied. It is believed that the mean particle size of the powder bed is predominant in influencing the granule formation mechanism, drop penetration time, and granule properties.

508017 Atomically Deposited Sintering Aids: Assessing the Effects of Alumina Particle ALD on the Sintering and Performance of SOFC Electrolytes and Dental Ceramics

Abstract:

The crystal structure of zirconium oxide (ZrO_2) can be tailored by the incorporation of aliovalent dopants, most commonly, yttrium oxide (Y_2O_3). The concentration of dopant dictates the thermodynamically stable crystal phase where 3 mol%
Y
conducting cubic phase. To achieve suitable mechanical properties and ionic conductivity, Y-TZP and YSZ ceramics must be near theoretical density, requiring high sintering temperatures. Previous researchers have successfully lowered the required Y-TZP and YSZ sintering temperatures through incorporation of aluminum oxide (Al_2O_3) particles. While this technique has proven to reduce the sintering temperature in both systems, the presence of particulate Al_2O_3 inclusions in densified parts reduces the homogeneity of the material which may have deleterious effects on densification and microstructure. In this work, we report on the use of atomic layer deposition (ALD) to conformally coat individual Y-TZP and YSZ particles with the desired amount of Al_2O_3. Constant-rate-of-heating (CRH) experiments were conducted using a horizontal push-rod dilatometer to extract the activation energy of sintering and also gain mechanistic insights into the active diffusion mechanisms as a function of ALD film thickness. Transmission electron microscopy (TEM) and scanning electron microscopy (SEM) were used to characterize the as-deposited thin Al_2O_3 films and the effects of the Al_2O_3 films on grain growth and microstructure during densification. Ionic conductivity and flexural strength tests were conducted to determine the effect of Al_2O_3 thin films on relevant material properties. Al_2O_3 incorporation by ALD is shown to have unique phenomenological effects on the densification of Y-TZP and YSZ as compared with conventional Al_2O_3 incorporation using particle-based approaches.

508644 Effect of Particle Size Distribution and Operating Parameters on Conduction and Convection Heat Transfer Mechanisms in Rotary Drums

Abstract:

Granular materials undergo process steps that include transportation, drying, heating and chemical or physical conversion. In several instances, these materials need to be heated or cooled during processing, and rotary drums (kilns) are the most commonly used process equipment for this purpose. It is not always possible to conduct experiments using different materials and consider all process conditions to study the thermal mechanism due to the inherent complexity of these processes. Therefore, computational modeling and simulations have been extensively used to understand the heat transfer mechanism in rotary drums. To maintain simplicity and to overcome the challenges of simulating polydispersed particles, a common practice is to assume that the polydispersed particles with a narrow size distribution behave like monodispersed particles thermally. However, no effort has been made to validate this assumption. This work focuses on understating the effect of polydispersed particles on the conduction and convection heat transfer mechanisms in indirectly heated rotary drums. It also aims to investigate the effect of rotation speed, fill level, particle size and their interaction effects on the conduction and convection heat transfer mechanisms in rotary drums. For this, the discrete element method (DEM) (using MFIX-DEM, an open-source computational fluid dynamics solver suite) simulation technique is used to analyze the thermal behavior of spherical silica beads, which are the most common catalyst support material.

507974 Selective Hydrogenation of Citral over Supported Pt Catalysts on Various Substrates

Abstract:

Highly dispersed platinum (Pt) nanoparticles (NPs) were deposited on various substrates by atomic layer deposition (ALD) for the reaction of citral hydrogenation. The substrates included multi-walled carbon nanotubes (MWCNTs), silica gel (SiO_2), commercial Î^{3}-Al_2O_3, and ALD-prepared porous Al_2O_3 particles (ALD-Al_2O_3). Based on TEM and CO-chemisorption analysis, the Pt NPs on all substrates were around 1.3 nm, and Pt/MWCNTs, Pt/ALD-Al_2O_3 and Pt/Î^{3}-Al_2O_3 showed similar Pt dispersion (75%-85%). But the Pt dispersion on SiO_2 was low (63%). In the reaction of citral hydrogenation, Pt/SiO_2 showed the highest activity due to the strong acidity of SiO_2 and the conversion of citral reached 82% after 12 hr with a selectivity of 58% of unsaturated alcohols (UA), including geraniol and nerol. Among all catalysts, Pt/MWCNTs showed the highest selectivity of UA (65%) due to its unique structure and electronic effect. In the cycling test, Pt/MWCNTs and Pt/ALD-Al_2O_3 catalysts were more stable than Pt/SiO_2 which could be attributed to the different interactions between the catalysts and the solvent.

508345 Numerical Simulation of Particle Classification in a Classifier Based on Coanda Effect

Abstract:

Efficient and precise classification of particles has been an important issue in many industrial applications as well as in academia. A relatively new particle classifier based on 'Coanda effect' was developed and reported to have many favorable features (Liang 2010). They include simultaneous multiple classification, simple design with no moving parts, and easy scale up to higher production rates. However, this classifier has attracted little attention so far, and our understanding of underlying mechanism associated with classification performance needs to be improved.

The purpose of this study is to investigate the detailed fluid flow and particle transport characteristics in a Coanda classifier using CFD (Computational Fluid Dynamics). The main focus is placed on the effects of working conditions (inlet flow velocity, particle diameter) and numerical method for turbulence dispersion of particles on the classification performance. As a numerical method, three different approaches are considered: k-ε model, k-ε model combined with DRW (Discrete Random Walk) model, LES (Large Eddy Simulation). In the first approach, standard k-ε model is adopted to obtain the mean flow and then particle trajectory is calculated based on the mean flow without consideration of turbulence dispersion. However, in the second approach using DRW model, the instantaneous flow velocity is constructed under the assumptions of isotropic turbulence in order to include turbulence effect. Finally, in LES, the instantaneous flow field is dynamically and accurately obtained at each time step. Numerical simulations are carried out for two-dimensional mean flow in case of k-ε model, but for three-dimensional unsteady flow in LES. Particles are tracked using a Lagrangian approach based on one-way coupling. We consider talc particles with density of 2,860 kg/m^3 and diameter of 1 to 20 μm.

The Coanda classifier is mainly composed of three inlet channels, three outlet channels, and Coanda block. Inlet channels are divided into one main inlet and two control inlets. The main air jet containing particles is supplied with the velocity of 35 m/s through the main inlet. Two control air jets with velocities of 32 m/s (lower) and 40 m/s (upper) are also supplied through the control inlets for enhanced control.

The trajectories of moving particles obtained from the mean flow (k-ε model) are analyzed for different diameters. The main jet containing particles deflects toward the curved Coanda block due to Coanda effect as expected. Fine particles ($d \le 2\mu m$) follow the flow streamlines and enter the lower outlet channel. However, coarse particles ($d \ge 8\mu m$) are detached from the streamlines and go into the upper outlet channel. This observation agrees well with the previous result (Liang, 2010).

Even though the mean flow and corresponding particle trajectories provide useful information, realistic particle passage do not take a form of single curve but they change in time due to flow turbulence. Therefore, particle trajectories are simulated again, turbulence dispersion being considered with DRW model. Due to turbulence dispersion, particles of the same diameter are scattered around the particle pathline obtained from the mean flow. As a result, particles of the same diameter can not be collected into a single outlet. For example, some particles (about 5%) of 1 μm enter the middle outlet even though most of them (about 95%) enter the lower outlet designed for collecting fine particles. This means that flow turbulence deteriorates the accuracy of classification.

Finally, LES simulations are performed to understand dynamic behaviors of fluid flow and particle transport. Instantaneous flow fields such as vorticity contours clearly show unsteady fluid motions inside the classifier. It is also shown that the generation of unsteady vortices leads to the scattering of particle trajectory.

A more detailed analysis including the classification efficiency will be given in the final presentation.

Acknowledgements

This research was supported by the National Research Foundation of Korea (NRF) under grant number 2017M2A8A4018482 and 2015R1D1A1A0105 9675.

References

Liang, S. (2010). *Chem. Eng. Comm.*, **197**, 1016-1032.

503491 Numerical Simulation of the Secondary Air Distribution Layer in the Gas Solid Fluidized Bed Based on Flunt Software

Abstract:

Due to the continuity beneficiation in the practical production, the gangue would accumulate on the distributor forming the gangue layer, which was called Secondary Air Distribution Layer and has vital effect on the gas distribution. In this paper, the ventilation characteristics of the secondary air distribution layer was analyzed. In addition, FLUENT software was used to study the effect of the modeled layer for the fluidity characteristics (e.g., pressure drop, bed height, volatility, etc.) in the gas-solid fluidized bed with the complex layer. The results showed that the complex layer could promote the secondary distribution of the airflow in the bed, which effectively inhibited the production of large bubbles and reduced the bed pressure drop fluctuation. Thus, itâ€™s of significance to provide synergy effect on the fluidization stability in the fluidized bed. Based on the above analysis, the introduced complex structure would be convenient to achieve the coal separation effectively.

507949 CFD-DEM Simulation of the Fluidization of Non-Spherical Particles in Fluidized Bed

Abstract:

In chemical and energy industries, fluidization concerning non-spherical particles (such as rod-like particle, disk-like particles, ellipsoidal particles, etc.) is often encountered, but research in this area is minimal. Yet understanding fluidization of non-spherical particles is crucial in many relevant industrial process. To investigate the fluidization of non-spherical particles, a CFD-DEM coupling algorithm is employed in this research to simulate a bubbling fluidized bed. The non-spherical particles are modeled by super-ellipsoids, multi-sphere models, and other particle models. The simulations results show that, the fluidization of non-spherical particles has significant differences from that of the spherical particles. The local void fraction and particle's projected face area perpendicular to fluid flow direction are directly related to the orientations of the non-spherical particles, which has a significant influence on the particle drag force and fluidization. Due to the significance of particle orientation, we further investigate the factors that influence the distribution of particle orientations. The results show that particle orientation depends on the fluidization time, fluidization velocity, and structure of the fluidized bed.

508320 CFD and DEM Simulation of the Cold Spray Process for Surface Coating with Fine Particles

Abstract:

The coating and functionalization of technical surfaces has been become increasingly important in the last two decades for various applications, since of its manifold possibilities for process optimization, as friction reduction, corrosion resistance or improved wetting behavior. Cold spray is a thermal spraying method for surface coating. It was developed in the late 1980s in Novosibirsk [1] and has gone through continuous optimization on certain applications since then. For example showed our previous study a reduction of the friction properties on steel pins, microstructured with fine Titania particles by an adapted spraying method [2]. In the cold spray process particles are dispersed in a gas stream, which is heated up to several hundred degrees and accelerated to supersonic velocities in a Laval nozzle. The particles impact on a substrate with about 800 m/s and form a strong bonding due to the high kinetic energy which is transferred to deformation and heat in the contact zone. In contrast to other thermal spraying methods, the temperatures occurring during the process are far below the melting temperatures of the used materials, which makes the process suitable also for thermal sensitive powders. The quality of the produced coating layers depends highly on the properties of the powder and substrate as well as on the nozzle design and the process parameters as initial temperature and particle velocities. Due to the high gradients, the high occurring velocities and the small size of the particles, typically in the range of 1 to 50 µm, the contact zone of impacting particles can only be measured indirectly and with destructive methods. Therefore, still not all phenomena which lead to a strong bonding in the contact area and a high quality coating are completely understood. For the study of these mechanisms two different simulation approaches have been used:

a) Simulation of particle and gas flow with Computational Fluid Dynamics (CFD)

For a better understanding of the spray process, the nozzle, the adjacent flow areas and the substrate were computed with a CFD method. The trajectories of single particles were calculated by an Euler-Lagrange approach. For the flow the k-Ï‰ SST model was used. Local temperatures in the substrate were calculated by the law of heat conduction. The impact conditions dependent on the nozzle geometry and the particle properties. With the local temperature distribution in the substrate and in the particle, precise impact conditions could be estimated. A study has been conducted to vary the parameters of the spray process, temperature and pressure as well as the particle properties as shape factor, density and size. Moreover the nozzle geometry has been varied in the simulation to optimize the spray process. The simulation results are compared to images of coated surfaces obtained by Scanning Electron Microscopy (SEM) for validation.

b) Simulation of particle-surface impact behavior with Discrete Element Method (DEM)

The DEM is widely used for investigations of particular processes. The advantage of this method is the detailed calculation of particle-particle or particle-wall contacts to obtain the important parameters as forces and deformations. The contact force in the particle wall contact is also dependent on the surface morphology. Therefore the surface topography of two real technical surfaces, a polished and an unpolished Titanium substrate were measured by Scanning Probe Microscopy (SPM) imaging with a nanoindenter (TI Premier, Hysitron). The surface topography was discretized and implemented in a DEM model. A parameter study has been conducted to determine the distribution of forces and contact angles occurring at the particle impact. Different particle sizes and impact velocities were studied.

The cold pray process dependents on various parameters which affect each other. Since of the high gradients and the small scales occurring during the process experimental measurements are often difficult. With this numerical study different parameters could be considered uncoupled of the other parameters. The results can be used for a better process understanding and an optimization of commercial cold spray systems.

[1] Alkhimov, A. P., Papyrin, A. N., Kosarev, V. F., Nesterovich, N. I., Shuspanov, M. M.: Gas dynamic spraying method for applying a coating; US Patent, 5302414; 1994

[2] Buhl, S., Schmidt K., Sappok D., Merz, R., Godard, C., Sauer, B., Kerscher, E., Kopnarski, M., Antonyuk, S., Ripperger, S.: Surface structuring of case hardened chain pins by cold-sprayed micro particles to modify friction and wear properties; Particuology 21, 2015

508568 Two-Fluid Validation of Constituive Models for the Simulation of Cylindrical Particles

Abstract:

Two-fluid simulations are a popular tool in the modeling of large processes. The constitutive models that close the governing equations are based in granular kinetic theory and typically assume that the particles are spherical. In reality, very few processes utilize spherical particles, hence there is a need for constitutive models that describe more complex shapes. Previous work used discrete element method (DEM) simulations to create a solid stress[1,2] and collisional dissipation rate[3,4] model for cylindrical particles that can replace the current spherical models. These new models have been added to MFiX and this work focuses on their validation. Cylindrical particles of various aspect ratios and solid densities were used in cratering and hopper discharge experiments. Simulations are designed to recreate the experiments in order to compare the results for crater depth and hopper discharge rate. The model effectiveness is assessed with regards to assumptions made concerning particle rotation and preferential alignment.

1. Guo Y, Wassgren C, Ketterhagen W, Hancock B, James B, Curtis J. A numerical study of granular shear flows of rod-like particles using the discrete element method. *J Fluid Mech*. 2012:1-26.

2. Berzi D, Thai-Quang N, Guo Y, Curtis J. Stresses and orientational order in shearing flows of granular liquid crystals. *Phys Rev E*. 2016;93(4):40901.

3. Buettner KE, Guo Y, Curtis JS. Using the Discrete Element Method to develop Collisional Dissipation Rate Models that Incorporate Particle Shape. *AIChE J*.

4. Berzi D, Thai-Quang N, Guo Y, Curtis J. Collisional dissipation rate in shearing flows of granular liquid crystals. *Phys Rev E*. 2017;95(5):50901.

509002 The Comparsion of Coarse Grained CFD-DEM for Simulating the Dense Bubbling Fluidized Bed" Requires Immediate Action

Abstract:

CFD-DEM is an effective tool to study the hydrodynamics in dense bubbling fluidized bed. However, the expansive cost of calculating the particles motion is not easy for its application in industrial application or dense particles system. Thus several coarse grained models (CGM) were put forward to simplifying the calculation of particles and decreasing the particles motion, such as PSM, SPA, MSPA,CGEE, EMMS-DPM where a group of particles are set as a parcel or coarse particle to reduce the computation. In this study the experiment and simulation test conducted to test five coarse grained methods. The result indicated that the models increasing the gravity force, pressure gradient force and drag force linearly with the coarse grained ratio (CGR) can be effective to ensure similar macro behavior between original system and coarse graining system. The simulation data of the solid fraction distribution and bed expansion ratio reflected good agreement as experiment. But the bubbles structure became less clear as increasing the CGR. Furthermore, the power spectrum density (PSD) and Wavelet analysis of the pressure signal were used and showed that SPA and EMMS-DPM gave better agreement result in term of dominant frequency and energy ratio at mesoscale feature. Finally the computational efficiency for different models was compared and the models of EMMS-DPM, PSM and CGEE have higher computation efficiency than SPA and MSPA relatively if the collision model of spring-damping were adopted.

509153 Stability Analysis of Uniform Gas Solids Flow

Abstract:

Uniform distribution of gas solids flow should be ensured to gain sufficient performance of fluidized beds. The highly chaotic entropy dissipation across solids bed will trigger non-uniform fluidization, although distributor plays an important role in obtaining homogenous gas stream. Hence, total solids pressure loading (Î¦$_T$), the pressure drop ratio between solids bed and distributor, is introduced here to present their interaction. The boundary between uniform and non-uniform fluidization can be detected by stability analysis based on Prigogineâ€™s minimum of entropy production principle, and this criterion for *N* parallel paths can be simplified into calculating the second derivative of pressure drop through one path with respect of Î¦$_T$. Furthermore, a phase diagram, illustrating the effects from operational parameter (Î¦$_T$), geometrical characteristics of distributor (C_d) and property of solids (*Ga*) on instability of uniformity, is attained to draw some useful conclusions of robust uniform fluidization.

509309 Numerical Simulation of Particle Sedimentation Related to Nuclear Safety By CFD-DEM Algorithm

Abstract:

Numerical Simulation of Particle Sedimentation related to Nuclear Safety by CFD-DEM Algorithm

Byoungcheol Hwang[1], Kiyofumi Moriyama[1], and Hyun Sun Park[1*]

[1]Division of Advanced Nuclear Engineering (DANE), POSTECH, Pohang, Korea

**corresponding author: hejsunny@postech.ac.kr*

Abstract

The Discrete Element Method (DEM) is a technique to treat particle-particle interactions, and is used in many industries such as civil engineering, powder processing, pharmacy, and so on. Coupling of this method with Computational Fluid Dynamics (CFD) is needed in problems where the particles interacts with fluid significantly, such as bubble column reactor, mineral processing, and so on.

The simulation of important phenomena in nuclear power plant accidents that involves significant core damage and melting, i.e. so called severe accidents, is also one of the topic where the CFD-DEM is useful. After the Fukushima accident in 2011, the severe accident analysis and preparation of counter measures to mitigate the accident consequences became regulatory requirements in many countries. One of the measures for the cooling and stabilization of the molten core is the flooding of the cavity under the reactor vessel where the dropping core melt is caught in a water pool. It is expected that the molten core breaks up into particles of several millimeter sizes and accumulates on the floor, then makes a particle bed, which is hopefully coolable. In this situation, the assessment for the coolability of the particle bed is very important, and a number of experimental and analytical approached suggest that the internal/external structure of the particle bed affects strongly the coolability. The CFD-DEM is a promising candidate for an analysis method that can clarify such details of the particle bed formation.

We tested the applicability of the multiphase CFD-DEM code, an open-source developed by CFDEM®coupling in 2011, for the simulation of particle sedimentation and particle bed formation under water. Our focus in the present work is on the prediction of the particle falling characteristics and particle bed shape. Although the real condition during a severe accident includes very high temperature particles (>1000K) and boiling water (two-phase flow) in a large scale, we started with a simulation of much simple experiment cases without heat transfer or gas phase, with spherical particles only as the first step for the validation of the method. The experimental data with solid particles of various diameters (4 mm, 6 mm) and densities (3600, 6000, 7800 kg/m^3), particle release nozzle diameters (20, 30, 40 mm) were referred to. The results of numerical simulation were compared with the experimental data in terms of particle falling time, maximum particle dispersion angle, particle average falling velocity, particle bed distribution, and so on. The particle falling time was well predicted by the simulation with the maximum deviation 5%. The maximum particle dispersion angle by the simulation significantly deviated from the experimental data probably due to the influences of initial disturbances. The particle average falling velocity was overestimated. The final particle bed profile by the simulation showed agreed with the experimental observation accurately in the peripheral region, but the height of the center part was underestimated. The fluid dynamic drag model and the particle-particle contact model would need improvement for better accuracy. The present work confirmed the validity of CFD-DEM algorithm on the application for the particle sedimentation in a simple liquid pool. The particle-pool interaction including the gas phase will be the next step.

Keywords: CFD-DEM; particle-fluid interaction; particle sedimentation; nuclear safety; severe accident

509440 Application of Extedned Discrete Element Method to the Melting Process of a Packed of Particles

Abstract:

Application of eXtedned Discrete Element Method to the melting process of a packed of particles

Mehdi Baniasadi, Maryam Baniasadi, Bernhard Peters, Xavier Besseron

Université du Luxembourg, Faculté des Sciences, de la Technologie et de la Communication, Luxembourg

e-mail: mehdi.baniasadi@uni.lu, maryam.baniasadi@uni.lu, bernhard.peters@uni.lu, xavier.besseron@uni.lu, web page: http:/www.xdem.de

ABSTRACT

The phenomena of melting takes place in many industrial processes, such as metal processing, environmental engineering and thermal energy storage systems. The problem of modeling of solid-liquid-gas flow including physical and chemical conversions of particles such as melting is the subject of intense research in industrial processes. These systems are difficult to model with numerical analysis due to the complex interaction between the different fluids and granular phases condering heat, mass and momentum excahnge. Basically models used to predict multiphase flows with particles are Eulerian-Eulerian and Lagrangian-Eulerian. The difference between the two models is based on different treatment of the movement of the particles. The particles are tracked individually in Lagrangian-Eulerian approach while they are considered as a continuum medium in Eulerian-Eulerian. Lagrangian-Eulerian is closer to the real physical processes rather than the Eulerian-Eulerian model for granular flows with particleâ€"particle interaction[1].

In this paper, our main focus is on the modelling of the melting process for a packed bed of particles subject to a fluid flow using a numerical technique based on Lagrangian-Eulerian so-called eXtended Discrete Elemnet Method (XDEM)[2]. In this approach, the particles are resolved as discrete phase coupled via heat, mass and momentum transfer to the surrounding fluid phase. The Discrete Element Method (DEM) coupled by heat transfer is used in order to evaluate temperature distribution, melting rate as well as velocity, oriantion and position for each particle individually. In addition the approach allows to be linked with continues numerical approaches such as Computational Fluid Dynamics (CFD) in order to solve fluid fields. The model has been validated by comparing predicted results with existing experimental data for melting of a single ice particle as well as a packed bed of ice particles in the horizontally flowing water. Fluid-particle and particle-particle intractions has also been considered. In addition, the model has the capability to be extended to the packed bed of particles with different size and properties to produce different liquid phases.

REFERENCES

[1] Dierich, F., Nikrityuk, P. A. & Ananiev, S. 2D modeling of moving particles with phase-change effect. *Chem. Eng. Sci.* **66,** 5459â€"5473 (2011).

[2] B. Peters, Thermal Conversion of Solid Fuels (WIT Press, Southampton, 2003).

514248 Real-Time Interpretation of Chaotic Characteristics and Instability of Dense Gas Solid Fluidization

Abstract:

The establishment of instability monitoring system targets at analyzing the chaotic characteristics and evaluating the instability of the fluidization process for dense gas solid fluidization. Experimental pressure fluctuation and pressure drop measurements were interpreted from dense gas solid fluidization system based on chaotic time-series analysis. The pressure fluctuation signals were filtered in terms of attenuating the noise signals. The filtered pressure signals were analyzed with respect to attractor reconstruction. Based on the reconstructed attractor, the maximum Lyapunov exponent and entropy were investigated and estimated under different bed heights and air velocities. Results show that the maximum Lyapunov exponents udner different bed heights and air velocities are positive, which indicates that dense gas solid fluidization exhibits chaotic characteristics. With the increase of bed height, estimated entropy of dense gas solid fluidization increases with the increase of bed height under identical air velocity, while it varies slightly with the increase of air velocity when the air velocity alters from minimum fluidization velocity to 1.36 times of the minimum fluidization velocity.

508510 Glucan Particles As Potential Carriers of Natural Flavonoids for Treatment of Idiopatic Inflammation Diseases

Abstract:

One of the novel methods of pharmaceutical formulation is the loading of the drug into biocompatible microparticles, preferably extracted from natural sources. This approach offers several potential advantages, *e.g.* increased bioavailability (the ratio of the administered drug, which actually enters the bloodstream) compared to conventional drug delivery methods, enabling the option of oral administration even for parenteral drugs. Multiple principles of the loading process are possible – drug encapsulation, adsorption onto the particle surface or precipitation inside the pores a hollow particle. The presented work focuses on precipitation of a drug from an organic solvent inside glucan particles. Glucan microparticles are ellipsoid porous shells, which are obtained from cell walls of *Saccharomyces cerevisiae* (baker´s yeast). The particle size is 2-5 micrometers. Cell walls of glucan particles are mainly composed of β-1,3-glucans. In order to prevent inflammatory reaction in organism, the yeast organelles are removed by a series of alkaline and acidic extraction steps. Subsequently, present lipopolysacharides are extracted using different organic solvents.
Lymphatic cells are mainly present in inflammation area in organism where they are responsible for progress of inflammation. Glucan particles (GPs) interact with lymphatic cell´s receptors on their surface and after that β-1,3-glucans are phagocytosed by the lymphatic cells. Therefore, GPs represent an effective oral delivery system for transporting of biological active substances. Curcumin is an anti-inflammatory flavonoid, which occurs in plants of family *Zingiberacaea.* In our project, curcumin was delivered to an inflammation in organism using GPs. Loading of curcumin into GPs was enabled via precipitation of curcumin from organic solvents. Curcumin dissolved in organic solvents was added to the GPs and through pores penetrated into cavity of glucan shell. Next, the solvent was removed and the curcumin was precipitated in glucan microparticles. GPs with the flavonoid inside were observed using confocal microscopy and SEM (scanning electron microscopy). The achieved content of curcumin was successfully analyzed by other methods, which can not be mentioned directly due to article in preparation. Furthermore, the curcumin-loaded GPs were tested *in vitro* on macrophage like cells and *in vivo* on laboratory mice in order to evaluate their effectivity. The obtained results show that these composites are promising materials for biomedical applications such as efficient treatment of inflammatory diseases (e.g. idiopatic inflammatory bowel disease, rheumatoic arthritis).

508601 Use of Advanced Imaging Techniques in Tablet Disintegration Study

Abstract:

Certified testing of tablet disintegration and dissolution provides us useful information. However, there are many more that are hidden due to techniques' restrictions. Such information is e.g. mechanism of disintegration, change of surface morphology or uniformity of the substances' distribution in the tablet. Therefore, there are many techniques used to obtain the remaining information such as Magnetic Resonance Imaging (MRI), Fourier Transform Infra-Red (FTIR) or Raman spectroscopy and microscopy, Terahertz spectroscopy, Scanning Electron Microscopy (SEM) and other.

Presented work uses tablet made by direct compression of binary mixture of ibuprofen and lactose (based on previous results there was 0.2 - 0.5 w/w of ibuprofen content used) pressed at different compaction pressures (low, medium and high). The tablets were subjected to several solutions and were studied by techniques that have insight into the tablet (MRI) and ones that can see the surface (FTIR and SEM).

MRI experiments provided us with several interesting and several expectable results. As the content of ibuprofen was increasing so was increasing the dissolution time and the wetted layer was decreasing. Similar results were observed when the compaction pressure was increased. When the tablets were submerged to dissolution limiting solutions (saturated lactose solution and solution saturated with both lactose and ibuprofen) the disintegration was severely limited and some tablets began to increase their volume. However, in case of 0.2 w/w ibuprofen tablets it didn't matter whether there was dissolution discriminating solution or not, it disintegrated anyway.

FTIR microscopy showed, that not only the amount, but also the shape and size of the particles from which the tablets were compacted can be determined. These findings were further supplemented by the dry state information obtained by Scanning Electron Microscopy after different dissolution times. Further the FTIR-ATR imaging technique helped to understand the basic mechanisms of the tablet disintegration and to study the dissolution and water penetration into the model ibuprofen-lactose tablets.

508632 Correlation Study between Liquid Penetration and Mechanical Properties of Pharmaceutical Tablets

Abstract:

Main phenomenon influencing treatment of patient in pharmacy is amount of active pharmaceutical ingredient dissolved in medium such as water or blood. Even though this isa key step to achieve desirable medical effect it is not the first one occurring.The first step is governed by surface properties such as wettability and roughness and the first contact point is followed by liquid penetration which is highly dependent not onlyon type of used liquid, ionstrength, liquid density but also on mechanical and structural properties of tablets such as porosity, structure strength or particle bonds.Through many analytical methods we can describe each of thesesteps individually. Aim of this study is to find proper correlation amongseveralmethods such as sessile drop method, texture analysis (TA) and magnetic resonance imaging (MRI) of tablet during dissolution.

For experiments binary mixture of tabletwas used. Tablets consist of lactose (hydrophilic) and ibuprofen (hydrophobic)with variable content (from 0.2 to 0.5 w/w ratio) and were also variable in compaction pressure used on them (from low â€“ 5 kN through 10kN as middle to 18 kN as high).Three different solutions were used according to their saturation with primary substances to prevent the tablet from dissolving. Surface properties were evaluated on contact angle and soaking rate of liquid, both measured by sessile drop method. Texture analysis was performed by recording the material resistance during dry and wet state of tablet. Then the data was processed and trends were obtained.Tablets during the MRI experiments were submerged into dissolution medium and the proceeding wetting front was measured. Later the images were processed via image analysis and trends were obtained.

Results obtained from surface experiments show that higher the ibuprofen content is the more hydrophobic surface appears to be and therefore the contact angle is increasing. Contact angle is independent on manufacturing pressure but soaking rate on the other hand depends strongly.MRI images show that this technique is not suitable for rapidly disintegration tablets, where the calculated error is within experiment conditions. On the other hand, tablet with prolonged dissolution are well measured. Results from MRI experiments also state that local saturation with substances supress dissolution of tablet. Texture analysis of tablets show that they undergo immediately structural weakening after being poured with solution independently on type of solution. Rate of this structural change mainly depends on solution's acidity. Thenthe tablet decreases its resistance in time which can be interpreted as relaxation rate. With increment of ibuprofen or saturation with primary substances this rate decrease, which means that dissolution is suppressed. Correlation between MRI and relaxation rate was done. Results shows that the correlation is mainly dependent on texture analysis measurement. On the other hand, correlation between MRI and soaking rate appears to be more tentative.

Experiment characterising surface, structural and liquid penetration properties were investigated by three different techniques. Obtained results were evaluated and put to mutual correlation. From liquid penetration point of interest there is correlation between MRI and contact angle method. Also, a correlation between MRI and texture analyser was found.

508642 In Situ Drug Amorphisation By Microwave Irradiation Stabilized By Mesoporous Silica

Abstract:

One of the common issues in the pharmaceutical industry is the poor solubility of the newly developed active pharmaceutical ingredients (APIs). The solubility might be significantly enhanced by changing the form of the API from crystalline to amorphous form. The amorphous state is characterized by lower thermodynamic stability and thus better dissolution properties leading to better bioavailability. In addition, the amorphous API in a dosage form (e.g. tablets) must also remain stable and not recrystallize for at least the shelf life of the drug product (usually 2 - 3 years). Various methods are being utilized to prepare and stabilize amorphous formulations in the industry, however they usually involve difficult formulation steps or suffer from poor efficiency, low stability or can be patent obstructed.

In this work a stable amorphous formulation is prepared simply using the standard wet and dry granulation process or direct tableting. Crystalline form of API is used during the initial process. The amorphous form is then prepared *in situ* and â€œon demandâ€ either during manufacturing, or by the end user by applying microwave radiation (MW).

To stabilize the API in an amorphous form after the MW, multi-structured mesoporous silica particles were added to the formulation. The particles were prepared using an emulsion synthesis of octylamine in tetraethyl orthosilicate. Using ethanol as an inhibitor instead of the commonly used nitric acid as a catalyst reduces the silica primary nano-particles nucleation rate. The formed nuclei grow larger and then self-assemble around the octylamine droplets and form hollow micro-particles. The prepared microparticles were characterized using the Scanning electron microscopy (SEM) and the Transition electron microscopy (TEM). To characterize the porous structure, nitrogen sorption (BET) analysis was conducted. The results clearly show the particle assemblies contain macropores (cavities), mesopores and micropores and possess very high specific surface area and also very large pore volume. Together with the thermal and chemical stability of silica these properties make the particles well suitable for preserving the melted API in an amorphous form after the MW.

As a proof of concept, initially powder mixtures and later simple tablets composed of Ibuprofen (model API), silica micro-particles and a filler excipient were experimentally investigated. The *in situ* amorphization of ibuprofen was conducted using a commercial microwave oven. The heating properties of the oven, i.e. the spatial distribution of microwave heating was characterized. Various microwave power input and varying composition of the mixtures (excipient/Ibuprofen ratio) were tested with respect to the melting point of the Ibuprofen. The change from the crystalline to amorphous structure due to MW heating was analyzed by x-ray diffraction (XRD), differential scanning calorimetry (DSC) and FTIR-ATR techniques. The dissolution profiles of ibuprofen from both MW untreated and treated tablets were measured using the High Performance Liquid Chromatography (HPLC).

Based on these initial results, fully formulated tablets were prepared with various API dosage strength using a tablet press. Same compression force and same dwell time was used for all of the tablets. The final tablet composition includes excipient capable of absorption of the MW, which ensures a uniform heating in the whole volume of the tablet. Next it contains the silica microparticles, which by capillary forces absorb the API once melted and stabilize it after cooling down.

Lastly, the tablets contain standard filler, lubricant and disintegrant excipients. After the MW application the tablets change their structure due to the melted API intake into the silica micro-particles, forming visible pores in the tablet structure. This has a notable impact on the dissolution and hardness of the resulting tablets. For the final tablet characterization SEM, Texture Analysis, FTIR and Magnetic Resonance Imaging were used.

508646 Manufacture of Personalised Medicines By API Printing on Porous Tablets

Abstract:

Personalised medicine offers an opportunity to produce tailor-made medicaments for patients. In order to realize this vision, it is necessary to find an alternative approach to the conventional tablet design and production. One that will replace the existing mass production of drugs offering only a few dose options of the active pharmaceutical ingredient (API), while at the same time it must meet the strict reproducibility and a defined relationship between composition and release rate of active substances requirements.

One of such approaches is to print the API on demand on a solid substrate directly when it becomes needed. Appropriate dosing of API and the possibility to even combine several APIs into a single medicament improves the variability of the dosages available for prescription and reduces the overall number of pills the patient has to take. The goal of this work is to find a simple, automated and cost-effective method of preparing personalised medicaments, which could take place directly in the pharmacy shops or at home.

The method in this work consists of a precise dosing of a dissolved API into an already prepared substrate â€“ porous placebo tablets with high sorption capacity. The tablets are prepared using direct compression in a tablet press from a simple physical mixture of a binder, filler, disintegrant and lubricant. As a filler, mesoporous silica particles were used. The high pore volume provides the free space for API sorption into the tablets and the small size of the mesopores ensure the API is deposited in an amorphous form.

The prepared particles used in the formulation were characterized using Transmission electron microscopy (TEM), Dynamic light scattering (DLS) and nitrogen sorption analysis (BET). The prepared tablets hardness and uniformity was characterized both prior and after loading. The tablets after loading were characterized using the X-ray powder diffraction and the Fourier transform infrared spectroscopy to determine the API form and deposition uniformity, followed by dissolution tests to confirm the applied dosage strength and to measure the release profile. So far, several BCS Class II APIs were successfully infused into the substrates using precise on-demand dosing. Impregnation of various doses of API were accurate. Later characterization revealed no crystalline API were present and the dissolution profiles were very satisfactory.

509441 Understanding Phase Transition of Acetaminophen in the Bulk and Surface of Acetaminophen

Abstract:

Pharmaceutical solids processing often lead to the generation of new phases in the bulk and on the surface of the processed drugs. Phase transformation of the API can potentially change the physiochemical properties and influence the drug performances including solubility, dissolution and stability. To detect these changes, analytical tools such as XRD are usually employed. However, these techniques remain suitable for bulk analysis. Whereas, phase transition on the surface remains hard to detect due to the lack of literature, researches in this area and the absence of surface analytical techniques. The aim of this work is to investigate different phases (pure crystalline, partially crystalline partially amorphous and amorphous) of Acetaminophen drug by combining bulk and surface analytical techniques. For bulk analysis, FTIR and XRD have been employed. The former shows no significant differences between Acetaminophen various phases. Whereas the latter confirmed the presence of different phases indicated by a remarkable variation in the absorbance intensity the peak width. Where, high and sharp peaks obtained for the pure crystalline Acetaminophen, fewer peaks observed for semi crystalline Acetaminophen and a single broad peak observed for the amorphous Acetaminophen. For the surface analysis, Nitrogen K near-edge X-ray absorption fine structure (NEXAFS) spectra of a series of NEXAFS indicates that the precise determination of IPs and Ïƒ* shape resonance energies is sensitive enough to distinguish between the different phases of Acetaminophen. DFT calculations provides information on the molecular orbitals involved during electronic excitation occurred as a result of NEXAFS experiment.

509599 Treatment of Cystinosis through Vitamin E Modified Silicone Hydrogel

Abstract:

Cystinosis is an orphan disease caused by a genetic mutation that leads to deposition of cystine crystals in many organs including cornea. Ophthalmic manifestation of the disease can be treated with hourly instillation of cysteamine eye drops. The hourly eye drop instillation is tedious to the patients leading to poor compliance and additionally, significant degradation of the drug occurs within one week of opening the bottle, which further complicates this delivery approach. Our work focuses on designing a contact lens to treat the disease with improved efficacy compared to eye drops. Our goal is to design a lens that is safe and that can deliver a daily therapeutic dose of cysteamine to the cornea while retaining drug stability. We show that cysteamine diffuses out rapidly from all lenses due to its small size. The rapid release of cysteamine and other ophthalmic drugs can be prolonged significantly by integration of vitamin E aggregates inside of the polymer matrix of silicone hydrogels. The increase in release duration increases as square of the volume fraction of the loaded vitamin E fraction which leads to high efficacy and suggests that the mechanism of attenuation in drug release is formation of nanosized barriers. The increase in duration depends on the type of the contact lens. The barriers are not effective in single-phase hydrogel lenses while the effect in the bi-phasic silicone hydrogels depends on the type of the contact lens. Vitamin E incorporation increases the release duration of both ACUVUE®OASYS® and ACUVUE® TruEye™ but the effect is more pronounced in TruEye™ likely due to the low solubility of vitamin E in the lens matrix and higher aspect ratio of the barriers. This result along with the high aspect ratio seen through SEM imaging in silicone hydrogels suggests that barriers could be forming at the interface of the silicone and hydrogel phases. The presence of vitamin E has an additional beneficial effect of reduction in the oxidation rates, likely due to a transport barrier between the oxygen diffusing through the silicone channels and drug located in the hydrogel phase. Based on this study, both Acuvue®OASYS® and ACUVUE® TruEye™ can be loaded with vitamin E to design a cysteamine eluting contact lenses for effective therapy of cystinosis. The lenses must be worn for about 4-5 hr. each day, which is less than the typical duration of daily-wear. Vitamin E loaded contact lenses could be very useful for treating many other ophthalmic diseases as well.

517683 Systems Integration for Dry Granulation Based Continuous Pharmaceutical Tableting

Abstract:

Over the past decade, the pharmaceutical industry has seen significant progress in the mechanistic understanding of relevant unit operations, availability of multiple sensor technologies for inline monitoring of critical quality attributes and advances in data analytics to support manufacturing decision-making. These have enabled a systematic realization of continuous manufacturing for pharmaceutical tablets using the direct compaction route. However, the limitations and difficulty in handling of some particulate materials provide the impetus for investigating granulation routes to the continuous manufacturing of tablets.

In this work, the integration of dry granulation into a continuous tableting line is investigated. Specifically, the Process Analytical Technology (PAT) tools required for inline and at-line monitoring of the tableting line, and the semi-mechanistic models for the roller compactor and the tablet press are discussed. While the Kawakita model is used for the tablet press, the roller compaction process model is consisting of an integration of material balance and actuator dynamics from Hsu et. al (2010) with the modified Johansonâ€™s model presented by Liu & Wassgren (2016). The PAT tools, mechanistic equations, and automation system are integrated for measurement validation and data accuracy using data reconciliation in a structured ISA-95 network using DeltaV. The experimental data is presented from the operation of the continuous dry granulation line-based pilot plant at Purdue University.

509248 Modeling Granular Material Mixing and Segregation Using a Multi-Scale Model

Abstract:

A multi-scale model is presented for predicting the magnitude and rate of powder mixing and segregation in blenders. The model combines diffusion and percolation correlations from small-scale discrete element method (DEM) simulations with advective flow field information from large-scale finite element method simulations. The multi-scale model predictions closely match results from both large-scale DEM-only simulations and experiments, but they are much more computationally efficient than DEM-only simulations. Therefore, in contrast to DEM, the multi-scale model is well suited for predicting blending in industrial-scale systems.

Modelling Granular Media using Dynamical Density Functional Theory

T. Hurst, B. Goddard & R. Ocone

Motivation

Granular media is ubiquitous in industrial and physical problems. One of the characteristic qualities of granular media at the microscopic level is inelastic hard body collisions, which drive macroscopic behaviour. Modelling granular media is a complex and interesting problem. Solid-like phenomena make it difficult to approximate a granular system with a continuum model, while chaotic behaviour and inelastic collapse in hard particle systems introduce modelling challenges on the microscopic scale. We investigate uses of dynamical density functional theory (DDFT), augmented with data gained from microscopic models, for granular media simulation.

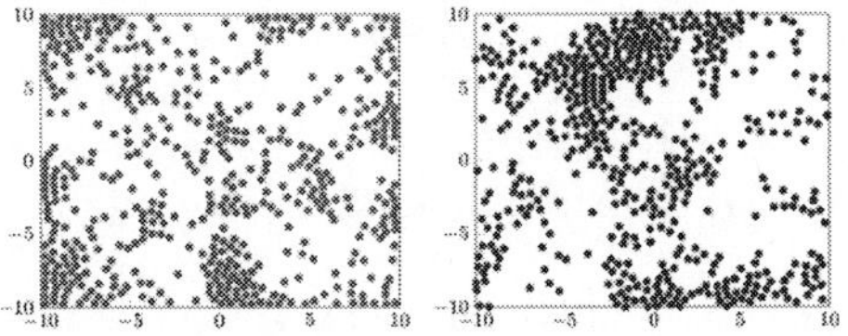

Figure 1: Inelastic systems with reflective (left) and periodic (right) boundaries

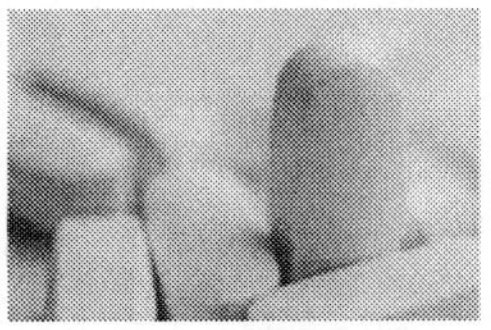

Dynamical Density Functional Theory

To derive a DDFT, we start with a set of Langevin equations for N hard particles of diameter σ in d dimensions with positions $\boldsymbol{r}^N$ and momenta $\boldsymbol{p}^N$:

$$\frac{d\boldsymbol{r}^N}{dt} = \frac{\boldsymbol{p}^N}{m}, \quad \frac{d\boldsymbol{p}^N}{dt} = -\nabla_{\boldsymbol{r}^N} V(\boldsymbol{r}^N, t) - \gamma \boldsymbol{p}^N + (mk_bT\gamma)^{1/2}\boldsymbol{G}(t), \tag{1}$$

on the hard sphere domain, $\mathcal{D}_N = \{(\boldsymbol{r}^N, \boldsymbol{p}^N) \in \mathbb{R}^{2dN} : \forall i, j, \|\boldsymbol{r}_i - \boldsymbol{r}_j\| \geq \sigma\}$, where $V(\boldsymbol{r}^N, t)$ includes external and interparticle potentials, and $\boldsymbol{G}(t)$ is an optional stochastic force term. The Kramer's equation is then a partial differential equation for its distribution function $f^{(N)}(\boldsymbol{r}^N, \boldsymbol{p}^N, t)$:

$$\frac{\partial f^{(N)}}{\partial t} + \frac{1}{m}\boldsymbol{p}^N \cdot \nabla_{\boldsymbol{r}^N} f^{(N)} + \nabla_{\boldsymbol{r}^N} V \cdot \nabla_{\boldsymbol{p}^N} f^{(N)} = \gamma \nabla_{\boldsymbol{p}^N} \cdot \boldsymbol{p}^N f^{(N)} + \gamma k_b T \nabla^2_{\boldsymbol{p}^N} f^{(N)}. \tag{2}$$

We integrate out all but one position and momenta in the Kramers equation, to find:

$$\frac{\partial}{\partial t} f^{(1)} = \mathcal{L}_1 f^{(1)} + \mathcal{L}^{2,\ldots,N}_{\text{part}}(f^{(k)}) + \mathcal{L}^{(1)}_{\text{diss}} f^{(1)} + \mathcal{L}_{\text{coll}}(f^{(2)}), \tag{3}$$

where $\mathcal{L}_1$ introduces effects from external potentials, $\mathcal{L}^{(1)}_{\text{diss}}$ effects due to dissipation and friction, $\mathcal{L}^{2,\ldots,N}_{\text{part}}$ effects from interparticle potentials, and

$$\mathcal{L}_{\text{coll}}(f^{(2)}) = -\sigma^{d-1} \int_{\mathbb{R}^d} d\boldsymbol{p}_1 \int_{\mathbb{R}^d} d\boldsymbol{p}_2 \int_{S_1^{d-1}} (\boldsymbol{p}_1 - \boldsymbol{p}_2) \cdot \omega f^{(2)}(\boldsymbol{r}_1, \boldsymbol{r}_1 + \sigma\omega, \boldsymbol{p}_1, \boldsymbol{p}_2, t) d\omega, \tag{4}$$

is a general collision term that arises due to the hard-sphere domain, when assuming only binary collisions. By taking the first two $\boldsymbol{p}$-moments and upon additional assumptions we arrive at a closed set of equations to model the dynamics of the system, a DDFT [1, 2]:

$$\frac{\partial}{\partial t}\rho = -\nabla \cdot (\rho \boldsymbol{v}), \tag{5}$$

$$\left(\frac{\partial \boldsymbol{v}}{\partial t} + \boldsymbol{v} \cdot \nabla \boldsymbol{v}\right) + \gamma \boldsymbol{v} = -\frac{1}{m}\nabla \frac{\delta \mathcal{F}[\rho]}{\delta \rho} - \sigma^{d-1} g_\sigma \int_{S_1^{d-1}} \rho(\boldsymbol{r} + \sigma\boldsymbol{\omega}/2)\rho(\boldsymbol{r} - \sigma\boldsymbol{\omega}/2)$$
$$\times \boldsymbol{\omega} \cdot \left\{\frac{k_B T}{m}\mathbf{1} + \boldsymbol{v}(\boldsymbol{r} + \sigma\boldsymbol{\omega}/2) \otimes \boldsymbol{v}(\boldsymbol{r} + \sigma\boldsymbol{\omega}) - \boldsymbol{v}(\boldsymbol{r} + \sigma\boldsymbol{\omega}/2) \otimes \boldsymbol{v}(\boldsymbol{r} - \sigma\boldsymbol{\omega}/2)\right\} \mathrm{d}\boldsymbol{\omega}, \tag{6}$$

where $\mathcal{F}$ is the well-studied Helmholtz free energy operator, which can include particle-particle interactions and external potentials, ρ is the density, $\boldsymbol{v}$ is the local velocity and g_σ is the value of the correlation function at σ, which must be determined. We choose to model g using statistical data from microscopic simulation.

DDFT Simulations

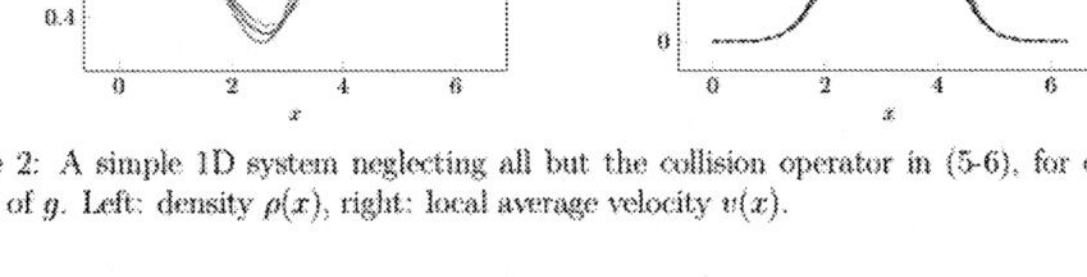

Figure 2: A simple 1D system neglecting all but the collision operator in (5-6), for different values of g. Left: density $\rho(x)$, right: local average velocity $v(x)$.

Data-Driven Modelling of g

A system of N hard spheres of diameter σ in d dimensions with positions $\boldsymbol{x}^N$ and velocities $\boldsymbol{v}^N$ obey

$$\frac{d\boldsymbol{r}^N}{dt} = \boldsymbol{p}^N, \quad \frac{d\boldsymbol{p}^N}{dt} = V(\boldsymbol{r}^N, t), \tag{7}$$

with a pairwise collision condition: if particles i and j are in contact, i.e. $\|\boldsymbol{r}_i - \boldsymbol{r}_j\| = \sigma$ and $(\boldsymbol{r}_i - \boldsymbol{r}_j) \cdot (\boldsymbol{p}_i - \boldsymbol{p}_j) < 0$, we instantaneously change the velocities via the rule:

$$\boldsymbol{p}_i^{out} = \boldsymbol{p}_i^{in} - \frac{1+\alpha}{2}\boldsymbol{\nu}_{ij} \cdot (\boldsymbol{p}_i^{in} - \boldsymbol{p}_j^{in})\boldsymbol{\nu}_{ij}, \quad \boldsymbol{p}_j^{out} = \boldsymbol{p}_j^{in} + \frac{1+\alpha}{2}\boldsymbol{\eta}_{ij} \cdot (\boldsymbol{p}_i^{in} - \boldsymbol{p}_j^{in})\boldsymbol{\eta}_{ij}, \tag{8}$$

where $\boldsymbol{\eta}_{ij} = \frac{\boldsymbol{r}_i - \boldsymbol{r}_j}{\|\boldsymbol{r}_i - \boldsymbol{r}_j\|}$ and $\alpha \in [0, 1]$ is the coefficient of restitution. For simple V, the Newton equations can be modelled using Event Driven Particle Dynamics (EDPD). The cell method and bounded increasing priority queues [3] reduce the computational effort to $\mathcal{O}(1)$ per collision. Using statistical methods for synthetic data (e.g. emulation [4]) we gain an understanding of the correlation function g near equilibrium, which we can include in (6-7).

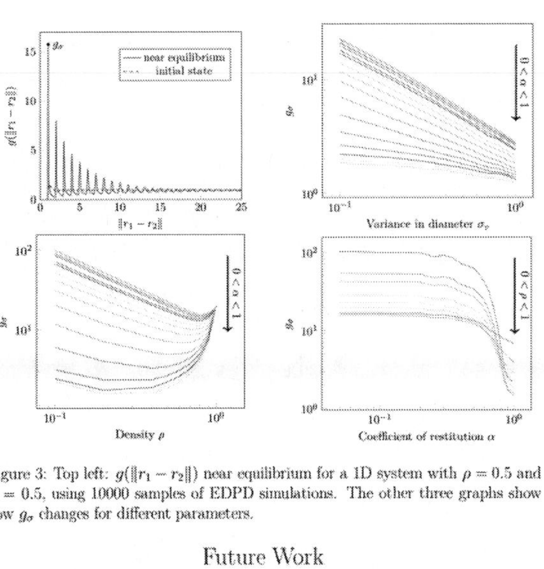

Figure 3: Top left: $g(\|r_1 - r_2\|)$ near equilibrium for a 1D system with $\rho = 0.5$ and $\alpha = 0.5$, using 10000 samples of EDPD simulations. The other three graphs show how g_σ changes for different parameters.

Future Work

We wish to include additional moments in the DDFT derivation, simulate DDFT in higher dimensions including interparticle potentials, and use EDPD in higher dimensions to calculate g.

[1] Archer, A., *J.Chem.Phys.*, **130**, 014509, (2009).
[2] Goddard, B. et. al., *Phys.Rev.Lett*, **109**, 1-5, (2009).
[3] Paul, G., *J.Comp.Phys.*, **221**, 615-625, (2011).
[4] O'Hagan, A., *Reliab.Eng.Syst.Saf.*, **91**, 1290-1300, (2006).

The support from EPSRC (grants EP/N034066/1, EP/L025159/1 and EP/L016508/01) is kindly acknowledged.

509555 A Study on Partially Wetted Particle Collisions with a Wet Wall

Abstract:

For the production of poly-ethylene, a gas-phase polymerization reaction is executed in a fluidized bed reactor. The polymerization reaction produces an excess of heat and the reactor is unable to efficiently remove the excess heat. The introduction of an Inert Cooling Agent (ICA) solves the temperature problem but also has a downside. Introducing the ICA in the reactor influences the particle collisions, leading to substantially different flow behavior. Besides the reduction of the restitution coefficient, the introduction of the liquid can result in adhesion instead of collision. This adhesion results in the formation agglomerates in the reactor and reduces the efficiency of the process. Due to the complex physics of this problem, Computational Fluid Dynamics (CFD) is a suitable method to gain insight in this phenomenon.

In order to study this complex system of partially wetted particle collisions a Direct Numerical Simulation (DNS) is used for incompressible flow. To model the coupling between the solid phases and the fluid phase, a second order accurate Immersed Boundary Method (IBM) is used (Deen et al.,2012). The interface between the fluids is tracked by a Volume Of Fluid (VOF) method (Baltussen et al., 2017). The contact-line dynamics are solved by imposing a contact angle as a boundary condition (Patel et al., 2017).

The simulated results will be compared with dry particle collisions on a wet wall (Tang et al., 2017) and with wet particle collisions on a wet wall.

References:

1. Deen, N.G., Kriebitzsch, S.H., van der Hoef, M.A. and Kuipers, J.A.M., Direct numerical simulation of flow and heat transfer in dense fluidâ€“particle systems. Chemical Engineering Science, 81 (2012): 329-344.

2. Baltussen, M. W., Segers, Q. I. E., Kuipers, J. A. M., and Deen, N. G., Cutting bubbles with a single wire. Chemical Engineering Science 157 (2017): 138-146.

3. Patel, H.V., Das, S., Kuipers, J.A.M., Padding, J.T. and Peters, E.A.J.F., A coupled Volume of Fluid and Immersed Boundary Method for simulating 3D multiphase flows with contact line dynamics in complex geometries. Chemical Engineering Science, 166 (2017): 28-41.

4. Tang, Y., Buck, B., Heinrich, S., Deen, N.G., Kuipers, J.A.M., Interface-resolved simulations of normal collisions of spheres on a wet surface. American Institute of Chemical Engineers, (2017).

509716 Water Expulsion from Carbon Rods at High Humidity (Invited)

Abstract:

Porous materials including carbons, silica gels and metal organic frameworks (MOFs) are used for water management in various applications including catalysis, dehumidification and heating, ventilation and air conditioning (HVAC). Materials performance has been enhanced through chemical modifications, surface treatments or impregnation with hygroscopic salts. Regardless of the sorption behavior mechanism, all hydrophilic porous materials exhibit an isotherm in which the quantity of water adsorbed increases with an increase in relative humidity. Recently, we have produced a new class of carbon based materials that exhibit unique water adsorption phenomena. The carbon-based rods can adsorb water at low humidity and spontaneously expel about half the adsorbed water at high relative humidity. Rods synthesized under different reaction conditions (e.g., reaction temperature, and the amount of $FeCl_3$ catalyst used) have different morphology, size and exhibited different RH for onset of water expulsion. We will present the physical and spectroscopic evidence for this behavior and its potential applications.[1]

References

1 Nune, S. K. *et al.* Anomalous water expulsion from carbon-based rods at high humidity. *Nat Nanotechnol* **11**, 791-797 (2016).

506317 Carbon Nanotubes to Immobilize Heavy Metals in Contaminated Soils (Invited)

Abstract:

Carbon Nanotubes to immobilize heavy metals in contaminated soils

António Alberto S. Correia[1], Martim P. S. R. Matos [2] and Maria G. Rasteiro[3]

[1] *Department of Civil Engineering and CIEPQPF – Research Centre for Chemical Processes and Forest Products, Univ. of Coimbra, Rua Sílvio Lima, 3030-788 Coimbra, Portugal*

[2] *MSc Student, Department of Civil Engineering, Univ. of Coimbra, Rua Sílvio Lima, 3030-788 Coimbra, Portugal*

[3] *Department of Chemical Engineering and CIEPQPF – Research Centre for Chemical Processes and Forest Products, Univ. of Coimbra, Rua Sílvio Lima, 3030-790 Coimbra, Portugal*

The contamination of soils with heavy metals is a growing concern in modern societies. When a soil is contaminated with heavy metals, which are inorganic, destruction of the contaminants is not an option since they are not biodegradable (Lone et al. 2008; Kroopnick 1994). To avoid the spread of contamination, soil stabilization techniques can be applied mixing materials with the soil in order to partially immobilize heavy metals. Carbon Nanotubes (CNTs) are nanomaterials known for its exceptional properties, like high surface area and adsorption capacity. Due to these unique properties, the potential use of CNTs in heavy metal contaminated water treatment has been studied, with very satisfactory results (Kabbashi et al. 2009; Abdel Salam 2013), however their application in contaminated soils is practically unexplored.

This experimental work is focused on studying the potential of using CNTs in soil remediation, especially to immobilize heavy metals. First, a large quantity of soil was collected and artificially contaminated with several heavy metals: lead (Pb2+), copper (Cu2+), nickel (Ni2+) and zinc (Zn2+), commonly present in contaminated soils. In order to avoid CNTs agglomeration, which originates the loss of their beneficial properties, an aqueous suspension of CNTs was prepared using a non-ionic or anionic surfactant combined with ultrasonic energy, to promote CNTs dispersion. Combination of two surfactants was also explored. Then, the soil, with and without the addition of CNTs, was subjected to adsorption tests to evaluate the CNTs capacity to improve heavy metals immobilization. To validate the adsorption test results, permeability tests were executed, simulating the conditions of a real case scenario.

The results obtained led to conclude that the addition of a small amount of dispersed CNTs can successfully increase the adsorption capacity of the soil and consequently improve the immobilization of heavy metals in the soil matrix (Matos 2016). The immobilization percentage varies with the type of heavy metals under study. For the soil studied, CNTs were most effective in the case of Ni2+ and Zn2+ immobilization, while in the case of Pb2+ and Cu2+ the addition of CNTs had almost no effect. Moreover, lixiviation tests conducted proved that CNTs were not released into water once they had been properly adsorbed in the soil.

Acknowledgement

The authors would like to acknowledge the financial support of Pest/C/EQB/UI0102/2013 from FCT/MCTES (PIDDAC), co-financed by the European Regional Development Fund (ERDF) through the program COMPETE (POFC).

References

Abdel Salam, M. 2013. 'Removal of heavy metal ions from aqueous solutions with multi-walled carbon nanotubes: Kinetic and thermodynamic studies', *International Journal of Environmental Science and Technology*, 10: 677-88.

Kabbashi, Nassereldeen A, Muataz A Atieh, Abdullah Al-Mamun, Mohamed ES Mirghami, MDZ Alam, and Noorahayu Yahya. 2009. 'Kinetic adsorption of application of carbon nanotubes for Pb(II) removal from aqueous solution', *Journal of Environmental Sciences*, 21: 539-44.

Kroopnick, P. M. 1994. "Vapor abatement cost analysis methodology for calculating life cycle costs for hydrocarbon vapor extracted during soil venting." In *Remediation of hazardous waste contaminated soils*, 779-90. RC Press.

Lone, Mohammad Iqbal, Zhen-li He, Peter J Stoffella, and Xiao-e Yang. 2008. "Phytoremediation of heavy metal polluted soils and water: Progresses and perspectives " In *Journal of Zhejiang University. Science. B*, 210-20. Hangzhou.

Matos, Martim. 2016. "Soil Decontamination using Nanomaterials." Msc Thesis, University of Coimbra, Coimbra, Portugal. (in Portuguese)

508522 Structuring and Functionalization of Iron Oxide Nanoparticles for Continuous Automated *in Situ* Protein Purification (Invited)

Abstract:

Magnetic nano- and microparticles have become highly promising in diverse areas of biomedicine, including *in vitro* applications in diagnostics and downstream processing due to their highly selective manipulation possible by applying external magnetic fields. For these applications, homogeneous particle properties are of crucial importance, allowing a uniform response and defined chemical and physical characteristics. In addition, the surface chemistry must be tailored for the intended application to ensure stability of the particles against agglomeration in the desired medium as well as the targeted interaction with the biological system.

We have established a multi-step process for the fabrication of biofunctionalized iron oxide nanoparticles, and studied their application for the purification of recombinant model proteins [1]. The surface chemistry of the particles was optimized by modification with a tailored ligand system capable of forming stable metal complexes with histidine (His), which is a standard strategy for the purification of recombinant proteins. The successful modification of the nanoparticles resulted in colloidal stability in aqueous biological media and could be systematically varied. The ligand is synthesized by coupling (3-glycidoxypropyl) trimethoxysilane (GLYMO) and Nî±,Nî±-bis(carboxymethyl)-L-lysine generating a molecule with a moiety structurally similar to nitrilotriacetic acid (NTA) which is well-known for its use in chromatography because it forms coordination compounds with His-tagged proteins via metal ions such as nickel. Hence the ligand system is denominated GNTA. Following, the addition of nickel ions resulted in certain agglomeration due to coordinate bonds with the NTA-moiety, facilitating the magnetic separation because of the increased total magnetic moment. Subsequently, extensive purification experiments were performed, utilizing recombinant proteins fused to His_6-tags produced by genetically engineered *Bacillus megaterium*. The protein binding capacity and separation efficiency of the nanoparticles were measured *ex situ* in a number of consecutive cycles of cultivation and compared to a commercially available product. The separation performance was analyzed utilizing enzyme-linked immunosorbent assay (ELISA) (protein activity), sodium dodecyl sulfate polyacrylamide gel electrophoresis (SDS-PAGE) (protein quantity), gravimetric analysis, elemental analysis and dynamic light scattering. Moreover, *in situ* purification was studied by adding the particles directly to the cultivation broth of *B. megaterium* in shaking flasks. By establishing a complex cycle of application and regeneration, we could successfully utilize the nanoparticles for separation of the recombinant model proteins using handheld magnets, achieving separation efficiencies of > 80 % with purities exceeding 99.5 % regarding protein components over five consecutive cycles. However, while the functionality of the ligand system could be successfully regenerated, there was a deficiency of stability of the particle agglomerates over several cycles due to the coordinate bond characteristic of the particle interactions.

In order to increase the efficiency of the magnetic separation of the particles over consecutive cycles, the as-synthesized superparamagnetic nanoparticles were structured in a spray drying process resulting in micrometer-sized aggregates. Thereby, the overall size of the magnetically separable structures is not dependent on the stability of the agglomeration state. Interestingly, the aggregates exhibit superparamagnetic-like behavior even in the micrometer range. It is assumed that organic molecules originating from the synthesis solvent are still bound to the surface of the particles preventing the fusion of the nanoparticles on the level of the crystal structure. The saturation magnetization is increased slightly and the total magnetization rises with the added mass of the densely packed aggregate compared with the agglomerates. Since the surface chemistry of the particles is not affected by the spray drying process, the developed functionalization method could be directly applied to the aggregates. An equal number of binding groups available for protein separation can thereby be provided, independent of the size of the used magnetic particles. The functionalized spray-dried aggregates were evaluated in a semi-automated *in situ* purification of the recombinant model protein directly from a growing cultivation of *B. megaterium* using a lab-scale stirred tank reactor with an external separation loop using handheld magnets. A separation efficiency of > 80 % with a purity exceeding 97.5 % regarding protein components could be achieved. The loss of magnetic particles over consecutive cycles could be significantly decreased in comparison with agglomerated particles. The set-up of the lab-scale bioreactor was further developed from a semi-automated into an automated mode [2]. Herein, the separation, washing, elution of the product, and regeneration of the functionalized particles were controlled using LabView. With this set-up, *in situ* purifications of the previously examined model protein and also an antibody fragment were performed. With an incubation time of the magnetic particles of 10 min and subsequent pumping of the cultivation medium through the external loop, 55 % of the protein and 83 % of the antibody fragment were separated from the cultivation medium. The purity of both products was determined to be > 97.5 %.

The preparation of spray-dried magnetic aggregates was also evaluated regarding the economical aspects. The applied nonaqueous synthesis of the highly crystalline iron oxide nanoparticles is cost-intensive due to the used precursors and synthesis method. Substituting a fraction of the iron oxide nanoparticles with silica particles in the spray drying process resulted in $SiO_2@Fe_xO_y$-core-shell aggregates based on the segregation process known from literature when using suspensions with particles of different size [3]. Since the shell is comprised of the smaller iron oxide particles, the functionalization procedure can be maintained. Additionally, the saturation magnetization of the resulting aggregates can be adjusted by varying the amount of silica particles offering. The total expenditure of the aggregate preparation could be decreased since the synthesis procedure of the silica particles is comparatively facile (Stöber process). Appropriate choice of iron oxide content would still provide aggregates with sufficient magnetization for the magnetic separation. Mechanical stability of the spray-dried aggregates was evaluated in consideration of the intended use in continuous production and separation of proteins in stirred tank bioreactors concomitant with mechanical stress from the stirrer in the reactor as well as pumps and pipe bends in the peripheral equipment. The particle size was determined before and after stressing with ultrasonication and no decrease of the mean size was observed. Additionally, nanoindentation measurements were conducted and the strength of the aggregates is such that they are expected to withstand any stresses that could occur in the intended application in a bioreactor. Here, especially the stability of microbial cells can be taken as reference since they must not be damaged in order to ensure further growth and synthesis of the desired product.

In conclusion, the presented approach allows facile and highly efficient selective purification of recombinant proteins and antibody fragments and appears promising for large-scale applications.

[1] J. Gädke, L. Kleinfeldt, C. Schubert, M. Rohde, R. Biedendieck, G. Garnweitner, and R. Krull, "In situ affinity purification of his-tagged protein A from Bacillus megaterium cultivation using recyclable superparamagnetic iron oxide nanoparticles," *J Biotechnol,* vol. 242, pp. 55-63, Jan 20 2017.

[2] J. Gädke, J.-W. Thies, L. Kleinfeldt, A. Kalinin, G. Starke, A. Lakowitz, R. Biedendieck, G. Garnweitner, A. Dietzel, and R. Krull, "Integrated in situ -purification of recombinant proteins from Bacillus megaterium cultivation using SPION in stirred tank reactors," *Biochemical Engineering Journal,* vol. 126, pp. 58-67, 2017.

[3] S. Zellmer, G. Garnweitner, T. Breinlinger, T. Kraft, and C. Schilde, "Hierarchical Structure Formation of Nanoparticulate Spray-Dried Composite Aggregates," *ACS Nano,* 2015/10/27 2015.

FUNCTIONALIZATION OF AEROGEL PARTICLES BY COATING IN A SPOUTED BED

Monika Goslinska[1], Ilka Selmer[2], Irina Smirnova[2], Stefan Heinrich[1]

1 Institute of Solids Process Engineering and Particle Technology, Hamburg University of Technology, Germany
2 Institute of Thermal Separation Processes, Hamburg University of Technology, Germany
corresponding author: monika.goslinska@tuhh.de

Abstract

Aerogels are porous and light-weighting materials with high specific surface area. Due to their properties they have been investigated as carriers of active ingredients. However, open pore structure limits their application because of changing their structure when they are exposed to liquids. Therefore, there is a protective layer on the surface of the aerogels needed, which preserves loaded drug or enables storage in liquid containing media. Furthermore, the knowledge of thickness of the applied coating layer is essential in order to predict the storage or release behavior.

In this work a novel technique for measurement of coating layer thickness of fine particles is presented. This method is based on cross-sectioning of single coated particles in micrometer size range with focused ion beam (FIB). This technique has been tested on two batches of produced aerogel particles coated with thin coating layers in a spouted bed.

Introduction

Aerogels are groups of highly porous solid materials with high inner surface area, open pore structure and low density. The combination of these extraordinary properties enables their application in different areas of life sciences. For example, due to high porosity and high adsorption capacity, the aerogels are used as matrix material for active compounds in food or pharmaceutical products [1]. These kind of particulate porous solids have to be coated with polymeric films for protective, functional or visual purposes [2]. The coatings can exemplary be used to stabilize the impregnated drug, to protect against influence of light, oxygen or water penetration during storage, to mask taste and odor, to improve the optical appearance or to design a defined drug release profile in the human body [2,3]. Especially in case of porous aerogels, the exposition of the porous network to water during storage or to gastrointestinal tract fluids during ingestion leads to change or destruction of their structure due to high capillary forces. Applying a protective, thin polymeric layer on the aerogel surface can avoid this problem and enables fulfilling the primary function of the particle [1].

Application of a thin film to a solid particle surface is a very complex process [4]. Usually, a coating material is sprayed on a particle collective. The success of the coating procedure depends on the used equipment or coating material type (melt, polymeric solution or dispersion) and other factors, like interactions between solid and sprayed liquid, process parameters and equipment parameters. The coating procedure may consist of the following simultaneously running steps: droplet formulation, wetting, spreading, agglomeration, evaporation and drying [5].

Thickness of the applied film is a very important quality parameter of the product. To predict shelf life of the product or encapsulated drug release profile is the knowledge about the layer thickness and its uniformity essential. There is a variety of layer thickness measurement methods, like mathematical calculations, micro tomography, X-ray, terahertz pulsed imaging etc. [6,7] which fail for the determination of the film thickness of fine aerogel particles because of the size of the particles. The search for new determination techniques of coating layer thickness is, therefore, an important and very challenging task.

In this work we show a new measuring technique of the thickness of a coating layer applied on a particle in a size of about 50 μm. For this we have combined usage of focused ion beam (FIB) for cross-sectioning of the particles and scanning electron microscope to image the cut thin film layer. The function of a FIB device is similar to that of a scanning electron microscope. The ion beam leads to excitation of secondary electrons, which are collected and transformed to an image. The resulting low quality images are used for positioning of the milling sections and monitoring of the milling procedure. This method enables milling of micrometer-sized material parts by integrated circuits, using a high current focused ion beam from a liquid gallium ion source [8]. FIB is frequently used for preparation of thin site-specific transmission electron microscope foils in materials science [9].

Materials and Methods

Aerogels

As a bed material already existing whey protein isolate (WPI) aerogels were used. The preparation process of these aerogel-microspheres can be found in [10]. For this study aerogel microspheres were produced at following conditions: pH = 7 of the gel, gelation at 80 °C, supercritical carbon dioxide drying at 120 bar and 40 °C for 8 h. The particles are characterized by the BET surface area of 207 m^2/g. Specific surface area of samples was measured using low temperature nitrogen adsorption–desorption (Nova3000e Surface Area Analyzer, Quantachrome Instruments, USA) and determined using the BET (Brunauer–Emmet–Teller) method. 20 mg of the samples were degassed under vacuum at 40 °C for 20 h before the analysis.

Coating solution

As a coating material 20 wt.-% shellac (SSB Pharma 56 Flakes, HARKE Pharma GmbH, Germany) solution in ethanol (≥99.8% with ca. 1 % MEK,Carl Roth GmbH, Germany) was used. To monitor the progress of coating procedure shellac solution was dyed with patent blue (Ruth GmbH & Co. KG, Germany).

Coating in spouted bed

Coating of the aerogel microspheres was performed in a in-house built spouted bed apparatus (see Figure 1). This experimental plant consists of a prismatic process chamber with two horizontal gas inlets, a conical expansion zone and a cylindrical relaxation zone. For the observation of the flow behavior two plexiglass plates were placed in the front and back sides of the process chamber. The fluidization air enters the process chamber from the bottom through two slots. The velocity of the gas can be adjusted by changing the height of these slots. The gas inlet flow rate was chosen to 4 m^3/h and increased up to 17 m3/h during coating of 80 g of particles. The temperature in the process chamber was adjusted to 30 °C. For the spraying of the coating solution a 2 mm two-fluid-nozzle was placed in the middle of the process chamber. Spraying of shellac solution with pressured air (1 bar) occurred in the direction of the gas flow in so called bottom-spray configuration. The coating solution was heated up to 35 °C and then sprayed using syringe pump with the flow rate of 0.5 ml/min. This very low spraying rate was chosen to avoid bed collapse during coating due to a too high amount of the liquid in the process chamber and in order to reduce additional drying time of the bed during operation. Particles were coated up to 340 min.

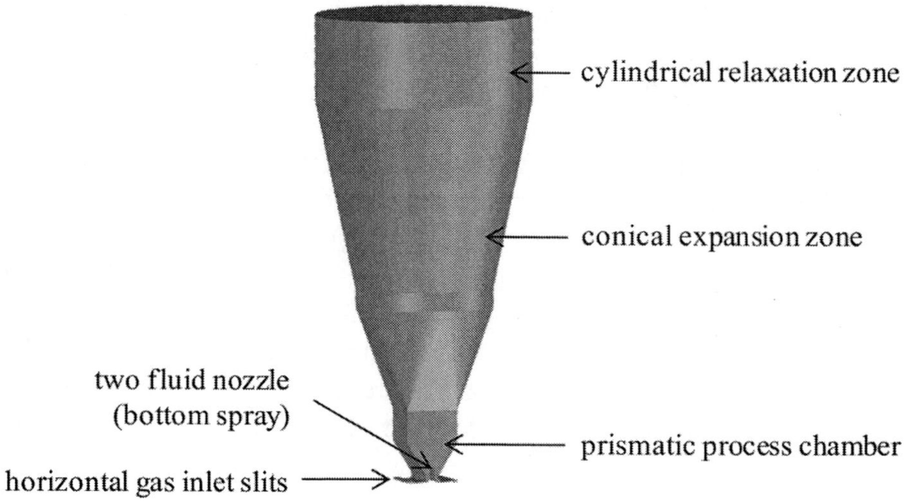

Figure 1: 3D scheme of the spouted bed apparatus.

Coating layer thickness

The layer thickness of a particle was determined by slicing using a combined SEM and FIB instrument (Helios NanoLab G3 DualBeam FEI, USA). To create the cross section of the single coated particle focused beam of gallium ions with high current of 2.50 nA was used. To make the cross section and the coating layer visible polishing with lower currents of 0.43 nA consecutively was performed. The measurement was performed for four particles with size below 45 μm in order to ensure that only single coated particles are present. Agglomerates could affect the visibility of the milling process and extend the milling process noticeable. Before milling the particles were sputtered with 7 nm gold layer in order to prevent charging with an electron beam. The milling process is shown in Figure 2 schematically. The SEM images were performed in order to measure the applied layer thickness at least at 10 different positions. The images were conducted at 2 kV and 50 pA.

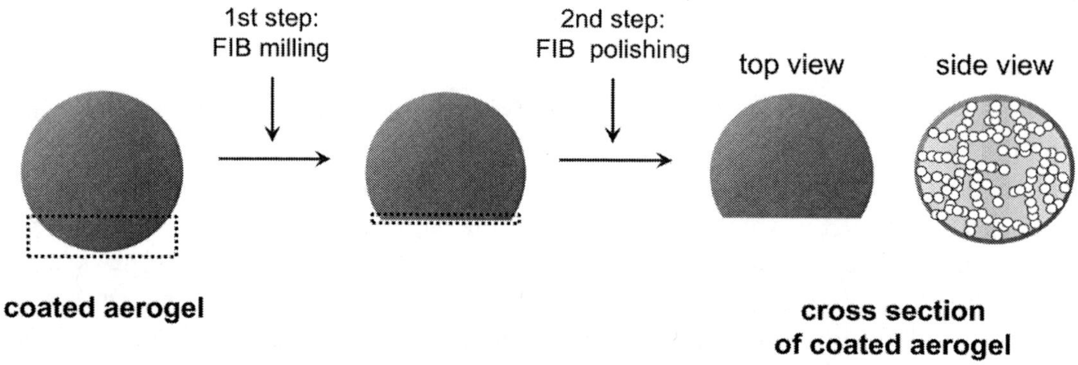

Figure 2: Scheme of FIB cross-sectioning process.

Results

The specific surface area of the produced particles is 207.0 m2/g and decreases to 19.0 m2/g for the coated particles. This refers to very good coverage of the pores and good coating uniformity for these particles, because nitrogen cannot penetrate inside the sample, so the accessible surface area drops. Therefore, the coated particles were cross-sectioned with FIB subsequently.

Figure 3 shows SEM images of a coated aerogel microparticle cross-sectioned with focused ion beam. The investigated particle has a mean diameter of 31.1 μm. In section 1 of the Figure 3 the top of the cross-sectioned particle is shown. The surface of the particle interacted with the ion beam. The interactions caused removal and damage of the applied coating layer, so that a very porous structure occurred. This is due to local heating and interactions of the sample with the ion beam. Hence, the coating layer thickness cannot be measured on the top of the cross sectioned particle. In section 2 of the Figure 3 the bottom section of cross sectioned particle is shown. In this part the applied shellac coating layer (bright) and the primary particle (dark) are shown. The layer thickness of this particle can be measured subsequently. The coating covers the particles surface at the marked section uniformly and does not penetrate or damage the particle deeply. For the investigated particle the average coating layer thickness was 675 ± 145 nm. The average coating layer thickness for four investigated particles was 750 ± 151 nm.

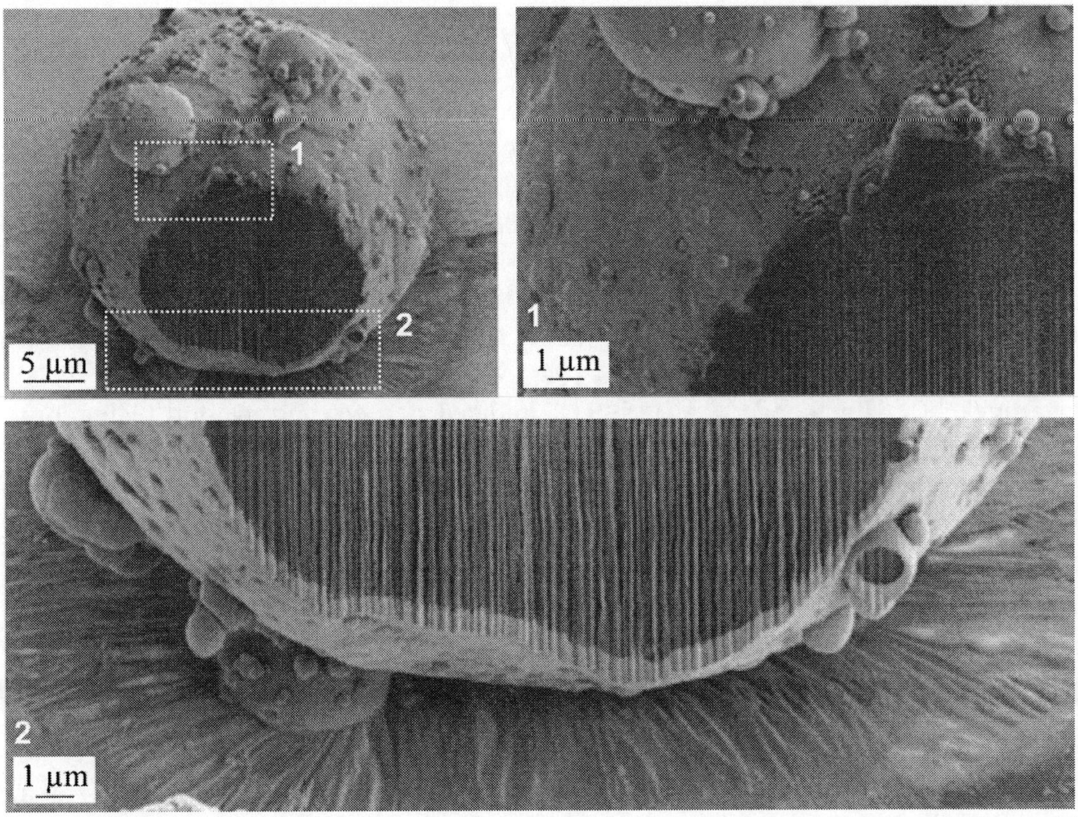

Figure 3: Scanning electron microscope images of an aerogel particle cross-sectioned with focused ion beam. Section A: damaged top of the sample due to interaction with ion beam during scanning. Section B: shellac coating layer (bright) in the bottom of the cross sectioned aerogel microparticle (dark).

Conclusions

In our work a novel measurement technique for coating layer thickness on microparticles, applied in a spouted bed process, was shown. The proposed method has an advantage over existing methods that it enables measuring of coating layer thickness of particles in micrometer size range very precisely.

In our investigation, the single coated particles were cross-sectioned using focused ion beam (FIB) coupled with SEM mode in two steps. In the first step, high ion beam current was used in order to remove a desired part of the particle up to the defined position. In the second step, low ion beam current was chosen to polish the pre-milled area and make the cross-section of the milled particle visible. The resulting layer thickness at the cross-sectioned part of the particles was clearly visible and was measured based on SEM images. The cross-sectioned particles showed the decrease of specific BET surface area up to 91 %. This is in an indication of uniformity of the applied coating. Additionally, the thickness of the film layers measured at FIB-cross-sectioned, single and coated particles is approximately 700 nm. The developed technique based on FIB cross-sectioning of coated particles enables successful measurement of coating layer thickness at any part of the coated microparticle.

References

1. Alnaief M., Antonyuk S., Hentzschel C.M., Leopold C.S., Heinrich S., Smirnova I. (2012). A novel process for coating of silica aerogel microspheres for controlled drug release applications. Microporous and Mesoporous Materials 160, 167–173.
2. Wen, H., Park, K. (2010). Oral Controlled Release Formulation Design and Drug Delivery: Theory to Practice. New Jersey: Wiley VCH
3. Gonçalves V.S.S., Gurikov P., Poejo J., Matias A.A, Heinrich S., Duarte C.M.M., Smirnova I. (2016). Alginate-based hybrid aerogel microparticles for mucosal drug delivery. European Journal of Pharmaceutics and Biopharmaceutics 107,160–170.
4. Dewettinck, K., Huyghebaert, A. (1999). Fluidized bed coating in food technology. Trends Food Sci Technol, 10,163-168.
5. Jones, D.M., (1985). Factors to consider in Fluid-bed processing. Pharmac. Technol., 9, 50-62
6. Czajkowska, M., Sznitowska, M., Kleinebudde, P. (2015). Determination of coating thickness of minitablets and pellets by dynamic image analysis. Int. J. Pharm., 495 (1), 347-353.
7. Markl, D., Zettl, M., Hannesschläger, G., Sacher, S., Leitner, M., Buchsbaum, A., Khinast, J.G. (2015). Calibration-free in-line monitoring of pellet coating processes via optical coherence tomography. Chem. Eng. Sci. ,125, 200-208.
8. Giannuzzi, L.A., Stevie, F.A. (2005). Introduction to focused ion beams – Instrumentation, theory, techniques and practice. New York: Springer Science+Business Media,
9. Wirth, R., (2009). Focused Ion Beam (FIB) combined with SEM and TEM: Advanced analytical tools for studies of chemical composition, microstructure and crystal structure in geomaterials on a nanometre scale. Chem. Geol., 261, 217-229.
10. Betz, M., García-González, C.A., Subrahmanyam, R., Smirnova, I., Kulozik, U. (2012) . Preparation of novel whey protein-based aerogels as drug carriers for life science applications. Sup. Flu., 72, 111-119.

508590 Investigating the Removal of Staphylococcus Aureus from Substrates By Modulating of Protein-Substrate Interactions (Invited)

Abstract:

Nosocomial pathogens can be transmitted through person to person, environment or contaminated water and food, infected individuals, contaminated healthcare personnel's skin or contact via shared items and surfaces. *Staphylococcus aureus* is the second leading cause of nosocomial bacteremia and also one of the most antibiotic resistance bacteria. *S.aureus* often interacts with the host via secreted and surface-anchored proteins, and is determined to be more strongly adhered to the host surface than E.coli.

The increase of multidrug-resistant strains has created an urgent need for the development of alternative approaches that focus on removal rather than killing of bacteria. To investigate the removal based approach, catatonically modified silica particles have shown promise in our laboratory. It has been hypothesized that the modified particle interaction with the bacteria must be stronger than bacteria- host surface interactions. In this study, a combination of silica particles and amino acids including arginine and tryptophan are expected to modulate the interactions between bacteria and surface. This knowledge will be helpful in developing additives for the suppression of protein-substrate interactions, thereby enhancing the removal of S.aureus and further reducing the transmission of nosocomial infections.

FABRICATION OF HIGHLY-FILLED COMPOSITES BY SPOUTED BED COATING AND STUDY OF THE INFLUENCE OF PARTICLE SHAPE ON MECHANICAL PROPERTIES OF THE MATERIALS

Eduard Eichner[1], Maksym Dosta[1], Stefan Heinrich[1], Gerold A. Schneider[2]

[1] Institute of Solids Process Engineering and Particle Technology

[2] Institute of Advanced Ceramics

Hamburg University of Technology, Hamburg, Germany

Abstract – In this work the mechanical properties, as modulus of elasticity, strength and fracture toughness, of copper-composite materials was studied. The volume fraction of copper was between 60 and 83 %. The study was carried out with two types of composites. The first type were composites containing spherical copper particles. The second type were composites with copper particles, which were rolled before pressing to a bulk material. The analysis of fracture toughness shows, that the composites with spherical particles have the highest fracture toughness of 1.58 $MPa{\cdot}m^{1/2}$. The composites with higher aspect ratios of particles have the highest fracture toughness of 2.64 $MPa{\cdot}m^{1/2}$. It corresponds to an increasing over 65 % for the highest values of the toughness referred to composites with spherical particles.

Introduction

In nature there exist many materials, which consist of two and more components and these materials are arranged in a particular way. Such materials are named composites. Composites can be artificial produced or created by the nature. Naturally occurring composites consist mostly of ceramics and polymers [1, 2]. Thereby, ceramic particles are uniformly distributed in a thin polymer matrix and the content of ceramics is significantly higher than that of the polymer. The high ceramic content in a polymer matrix is important for the good reinforcement of the polymer. But also the polymer is a very important part of a composite. The polymer is responsible for the ductility of the materials. One of the best investigated natural composite materials is nacre. Nacre consists of small ceramic platelets, which are surrounded by a polymer matrix (see fig. 1). Although nacre consists of brittle ceramic and soft polymer, nacre has good combination of mechanical properties like stiffness, strength and especially toughness. The good mechanical properties could be achieved by arrangement of particles and polymer in an intricate way [3]. On the one hand ceramic content in nacre is about 95 vol.%. Otherwise ceramic particles have a high aspect ratio and are organised in several hierarchical levels. Due to such a structure nacre has good mechanical properties.

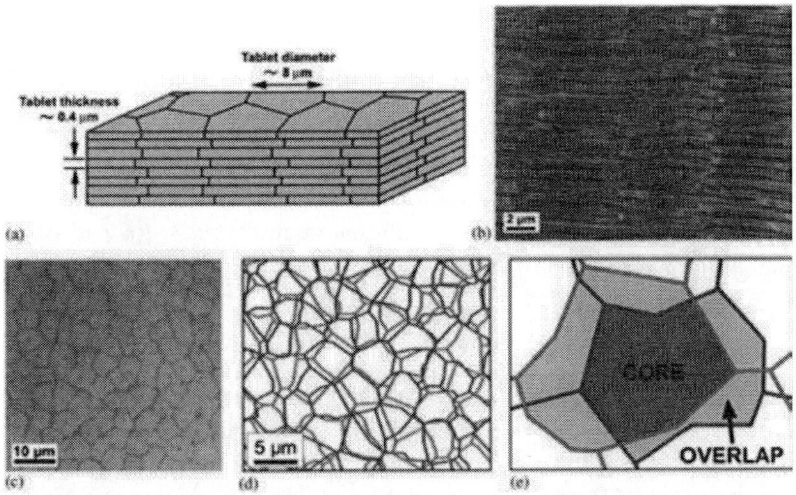

Fig. 1. Structure of nacre. (a) Schematic of the tablets arrangement in nacre; (b) scanning electron micrograph of a fracture surface in nacre; (c) top view of tablet tiling in nacre; (d) reconstitution of the arrangement of the tablets from one layer to the next; (e) core and overlap areas in the tablet arrangements [4].

Beside nacre more natural composite materials with hierarchical structure are known. These are for example elk antlers and toucan beak [5]. The natural composite material with highest known tensile strength are limpet teeth. The strength of the teeth is 3-6.5 GPa and also the modulus of elasticity with 120 GPa is very high [6, 7]. The good and sometimes extremely good mechanical properties of natural materials in spite of relative weak components gave an impulse to reproduce the natural structures. Therefore, in recently years, many attempts have been made to reconstruct the structure of the natural material and to achieve good mechanical properties. There exist some possible process routes to fabricate composite materials, but not every route can be applied for hierarchical materials. The problem in fabrication of hierarchical materials could be partially destroying of the existing hierarchical level during the build-up of the next level. This can happen due to the choice of an inappropriate solvent for example. One possible route for the reproduction of natural hierarchical composite materials is the spouted bed spray granulation process. The usage of the process has allowed achieving of high ceramic contents in composite materials [8] as well as reproducing of the hierarchical structure [9, 10].

Spouted bed is a special form of fluidized bed with many applications in particle technology, such as drying, granulation, coating and chemical reactions. Spouted bed can be used for different type of particles. The spouted bed technology was originally investigated as alternative to fluidized beds for operating of coarse particles [11]. A fluidized bed has a process chamber with a constant cross section. Therefore, all particles are fluidized at the same time. A spouted bed has, by contrast to the fluidized bed, an increasing cross section area of the process chamber. The form of the process chamber in a spouted bed can be conical as well as prismatic. Process air comes through a hole in the middle of the cone or through two slits in the chamber. As result only particles in the middle of the chamber are in the spouted state. So the energy consumption is significantly lower thereby. This is especially important for the application of coarse particles. Nevertheless, nowadays the spouted bed technology has also found application in processing of very fine particles. Spouted beds are used also for cohesive and nonspherical particles [11]. For processing of fine and cohesive particles a special spouted bed was designed. As it was mentioned above, it has been already successfully applied for fabrication of highly-filled and hierarchical ceramic-polymer composites [8, 9].

Beside of the high-filling degrees and hierarchical structure of composite materials, high aspect ratio plays an important role for good mechanical properties of composites. This can be seen in the structure of natural occurring composite materials [5]. Additional to the structure of natural composites Gao [1] postulates a theory about the influence of the aspect ratio on mechanical properties.

In this contribution we study the influence of the aspect ratio on mechanical properties of composite materials. The spouted bed was used for coating of particles with mean size of 29.5 µm, which are still outside the typical application range of conventional spouted beds. Instead of ceramics or pre-structured particles, copper particles were used.

Materials and Methods

For the fabrication of composites, copper particles with a mean diameter of d_P = 29.5 µm (Womet GmbH, Germany) were used as filler. The particles have a density of about 8.9 g/cm³ and a modulus of elasticity between 100 and 130 GPa. The particle size was measured with a Camsizer XT (Retsch Technology). The matrix polymer was polyvinyl butyral (PVB) Mowital B 30H, which was provided by Kuraray Europe. The polymer has an elastic modulus of approx. 2.5 GPa and a tensile strength of 34.8 MPa. PVB is soluble in ethanol and has a very good adhesion to hydrophilic surfaces. The complete approach for the fabrication of composite materials contains spouted bed spray granulation, hot pressing of coated particles and characterisation of mechanical properties and composition. The detailed description of the approach was done by [12]. But for the study the influence of the aspect ratio on composite materials properties, a further process step had to be added to the approach. The particles, which are coated in the spouted bed, have a spherical shape (aspect ratio about 1). For achieving of higher aspect ratios, particles had to change their form. By the additional step "rolling", the form could be changed from spherical one to platelet-like structure.

Before processing in the spouted bed polymer solutions were prepared. For the preparation, PVB in quantities between 10 g and 25 g was dissolved in ethanol with a constant concentration of 4 wt.%. In the first step the ethanol was heated up to a temperature of 30 °C and PVB was slowly added under stirring (magnetic stirrer, Witeg, Germany). For spraying-in of the solutions into the spouted bed a two-fluid nozzle (model 970, form 3 S1, Schlick, Germany) in the bottom-spray configuration and a peristaltic pump (TU 200, Medorex, Germany) were used. A fill consisting of 500 g of copper particles was used in each experiment. Figure 2 shows the geometry of the used spouted bed apparatus.

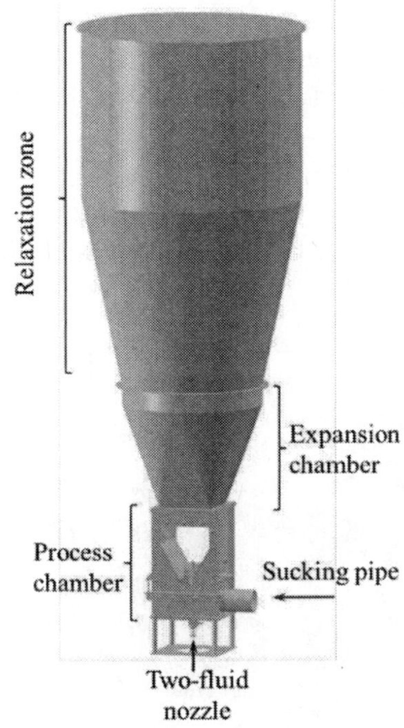

Fig. 2. Spouted bed for coating of fine particles.

This geometry is especially designed for the spouting of fine particles. The apparatus consists of a relatively small prismatic process chamber with two adjustable gas inlets, two conical parts and a cylindrical part. The bed is operated at under-pressure conditions. The air enters the apparatus through a sucking pipe, where it is heated up to the required temperature. From the pipe the air attains a wind box and then enters the prismatic process chamber through two thin horizontal slits with adjustable heights. For a given air volume flow, the height of the slits specifies the gas inlet velocity. In the process chamber two air streams change the flow direction from horizontal to vertical one by a profile in the middle of the process chamber and unite to a single gas jet over the profile. The gas transfers momentum to the particles and they follow the air within a turbulent spout. Particles carried by the gas jet pass the conical spraying zone. In the spraying zone, particles are coated by the polymer. The polymer is sprayed-in as a solution by the two-fluid nozzle. After coating, the particles enter the conical expansion zone and the cylindrical relaxation zone. In these zones, the gas velocity is significantly lower. The particles become decelerated, change finally their flow direction under the influence of the gravity force and slide downwards along the apparatus wall back in the process chamber. Hence the particle circulates in the apparatus, which is a characteristic attribute of the spouting process. During the period between leaving the spraying zone and the re-entry into the process chamber, the solvent from the polymer solution has to be evaporated in order to avoid agglomeration, resulting in formation of large lumps.

For the usage of fine particles in a spouted bed, it has to be properly designed. The process chamber should be relatively small in comparison to large expansion and relaxation zones. The expansion and relaxation zones have to be tall, because the strongly accelerated fine particles need a large height to be decelerated and to change the moving direction under the influence of the gravity. In this way elutriation of particles from the apparatus can be avoided (or at least strongly reduced). The major parameters for stable processing of fine particles are the air flow rate and the processing temperature as well as the polymer content in the solution and the spraying rate of the polymer solution. Air flow rate has to be high enough for a stable spouting, on the other hand it should not be too high, to avoid elutriation. Processing temperature, polymer content and spraying rate are significant for evaporation of solvent from the solution and for the occurring of agglomeration. The appropriate parameter range should be found individually for each particle-solution system. For the used system the temperature in the prismatic process chamber was controlled and adjusted on 30 to

40 °C. The volume flow of the spouting air was set to 16-20 m³/h at the beginning of the process. At the end of the process, the flow was increased up to 30-35 m³/h due to agglomeration of particles, which could not be completely avoided. The spraying rate of the solution was set to 5-10 ml/min.

For the measurement of mechanical properties of composites with an aspect ratio about one, the granulated and partially agglomerated particles were hot-pressed directly after the granulation process. For increasing the aspect ratio particles were rolled in a rolling mill. For it a rolling mill W40 (Dinkel Maschinenbau, Germany) was used. After the rolling, small platelets were formed. The rolling step is shown in figure 3.

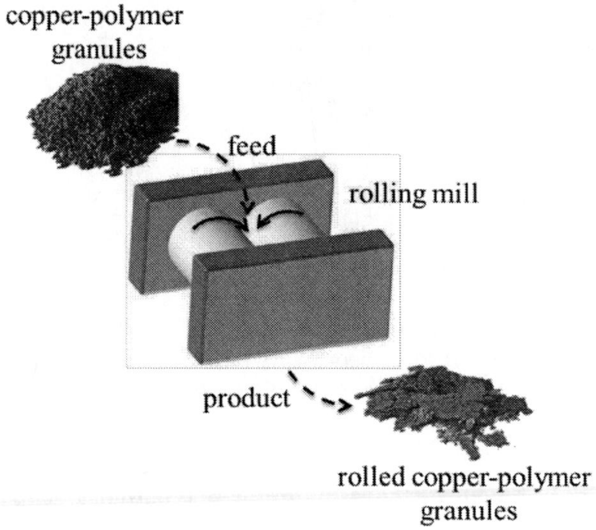

Fig. 3. Rolling step of coated copper particles.

The platelets were aligned in a press die and were pressed to tablets using an automatic press PWV 300 (Paul-Otto Weber GmbH, Germany). The temperature of the die was increased by a rate of 10 K/min to a final temperature of 160 °C at a small pressure of 40 MPa. At this temperature particles were pressed for 1 hour under isotherm conditions and at a pressure of 500 MPa. Not rolled particles were pressed directly after coating under same conditions. After pressing, the compact was grinded to get a smooth surface. The fabricated composite was cut in specimens with a rectangular cross section (size approximately 2-4 x 2 x 40 mm³ (height x width x length)) with a diamond saw Brillant 200 (ATM GmbH, Germany).

After pressing, the composition of the bulk materials was determined by measuring the geometrical density of the sample and by thermogravimetric analyses (TGA). For TGA 100 - 200 mg of the material m_{samp} were heated to the temperature of 450 °C (temperature higher than the decomposition temperature of PVB) and kept isotherm for 90 minutes. Nitrogen atmosphere was used in order to avoid an oxidation of copper at high temperatures. The complete calculation of the sample compositions can be found in (12).

Strength and modulus of elasticity of the fabricated composites were measured via four-point bending tests. Fracture toughness of the fabricated composite materials was studied by the measurement of crack resistance curve (R-curves) in load-displacement tests. For the measurements, three-point bending tests were carried out in a highly stiff loading device.

Results and Discussion

The mechanical properties were tested for different copper-polymer compositions. The results of the measured modulus of elasticity and strength are presented in figure 4. The results of the fracture toughness are presented in figure 5.

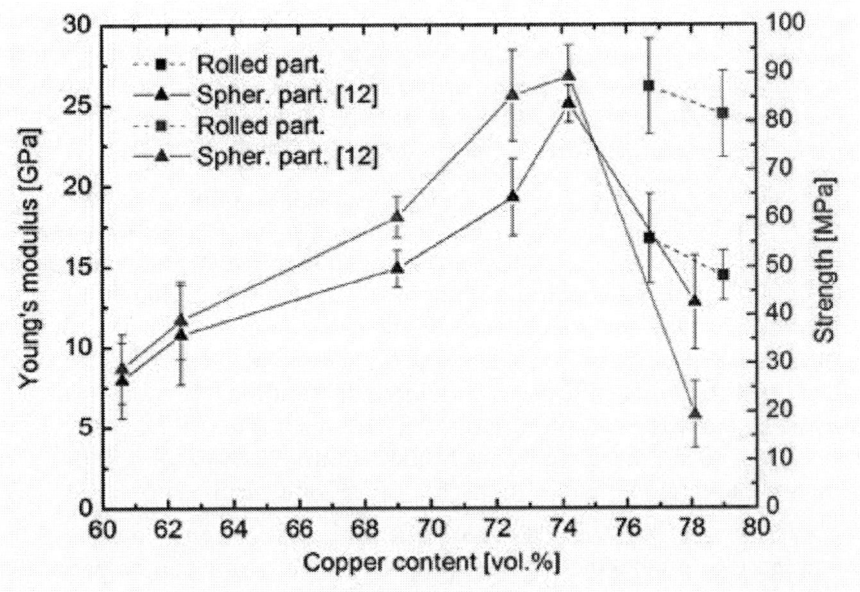

Fig. 4. Modulus of elasticity and strength of copper-polymer composites.

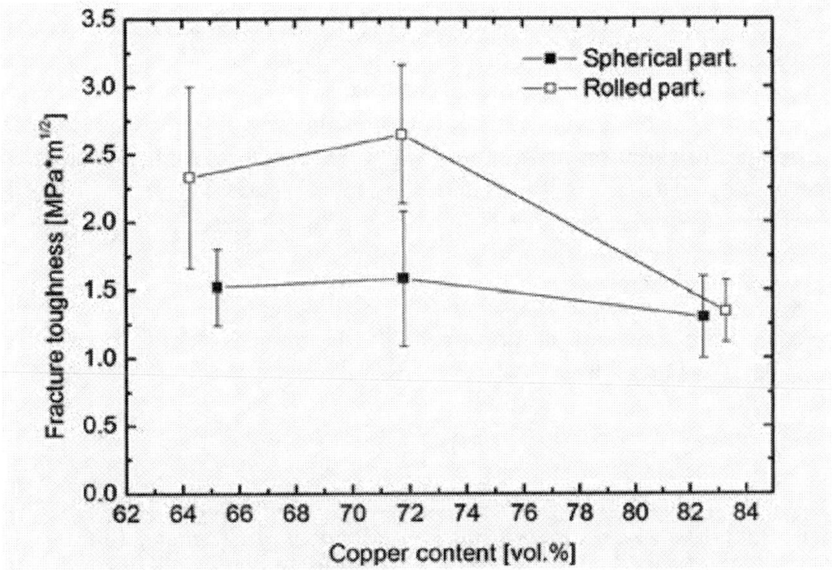

Fig. 5. Fracture toughness of copper-polymer composites.

Figure 4 shows the comparison of the mechanical properties for spherical particles, which were measured in a previous work [12], and rolled particles with high filling degrees of copper. The comparison shows that Young's modulus has approximatively the same value before and after the rolling. But the bending strength could be clearly raised by the rolling. Figure 5 shows that the composites with spherical particles have lower fracture toughness than composites containing particles with increased aspect ratio. The composites with spherical particles have relatively constant fracture toughness. At 65 vol.% of copper the fracture toughness value is 1.52 $MPa \cdot m^{1/2}$. At 72 vol.% of copper the value increase to 1.58 $MPa \cdot m^{1/2}$ and decrease again at 82 vol.% of copper to a value of 1.3 $MPa \cdot m^{1/2}$. In composites with rolled particles the fracture toughness is 2.3 $MPa \cdot m^{1/2}$ at 64 vol.% of copper. The value increases to 2.6 $MPa \cdot m^{1/2}$ at 72 vol.% and decrease to 1.33 $MPa \cdot m^{1/2}$ at 83 vol.% of copper. The measurements show a relative strong variation of fracture toughness values. The reason for this can be a relatively broad particle size distribution. During the crack propagation, the crack front is redirected and has to go around the particle. The surface of the crack, related to the crack growth, increases. The increase of the surface requires additional energy for the crack growth. So the elongation of the crack path increases the fracture toughness (13). If a crack has to bypass a relatively fine particle, the surface increase and fracture toughness are relatively small. On the other hand, if a crack has to bypass a larger particle, the increase of surface is higher and therefore fracture toughness also grows.

In composites containing particles with increased aspect ratio, particles were aligned perpendicularly to the notch. For this reason, the crack path and fracture toughness increase significantly. But also the difference in crack path elongation between fine and coarse particles is stronger. The aspect ratio of particle with the same size can also vary. Although the particles were uniformly fed to the rolling mill, the slit between the rolls of the mills stayed constant and the rolls were rotated approximatively with a constant velocity, the amount and the size of the particles, which come at the same time in the mill, vary. The force distribution among the particles in the mill is changed and particles are deformed differently. So the aspect ratio of the particles can also vary with the same size. Due to the described possibilities a stronger variation of fracture toughness values in composites containing particles with higher aspect ratios can be observed.

At copper filling degrees over 80 vol.% polymer fraction is to low and composite becomes very weak. Therefore, at these filling degrees rolling does not play any role and composites have almost the same fracture toughness.

Conclusions

In this work the influence of aspect ratio on mechanical properties was studied. Copper-polymer composites containing spherical particles (aspect ratio ~1) and containing particles with higher aspect ratios were produced and their mechanical properties were determined via bending tests. The results show that composites containing particles with higher aspect ratios have better mechanical properties. Even though the variation of fracture toughness values became stronger. The toughness of composites containing particles with increased aspect ratios is over 50 % higher than in composites with spherical particles between 62 and 75 vol.% of copper.

Acknowledgements

We gratefully acknowledge financial support from the German Research Foundation (DFG) via SFB 986 ''M3'', projects A3 and A6.

References

[1] Gao, H. (2006). Application of Fracture Mechanics Concepts to Hierarchical Biomechanics of Bone and Bone-like Materials. *International Journal of Fracture*, *138*(1-4), 101–137.

[2] Barthelat, F., & Espinosa, H. D. (2007). An Experimental Investigation of Deformation and Fracture of Nacre–Mother of Pearl. *Experimental Mechanics*, *47*(3), 311–324.

[3] Barthelat, F. (2007). Biomimetics for next generation materials. *Philosophical transactions. Series A, Mathematical, physical, and engineering sciences*, *365*(1861), 2907–2919.

[4] Barthelat, F., Tang, H., Zavattieri, P. D., Li, C.-M., & Espinosa, H. D. (2007). On the mechanics of mother-of-pearl: A key feature in the material hierarchical structure. *Journal of the Mechanics and Physics of Solids*, *55*(2), 306–337.

[5] Chen, P.-Y., Lin, A. Y. M., Lin, Y.-S., Seki, Y., Stokes, A. G., Peyras, J., Olevsky, E. A., Meyers, M. A., & McKittrick, J. (2008). Structure and mechanical properties of selected biological materials. *Journal of the mechanical behavior of biomedical materials*, *1*(3), 208–226.

[6] Barber, A. H., Lu, D., & Pugno, N. M. (2015). Extreme strength observed in limpet teeth. *Journal of the Royal Society, Interface*, *12*(105).

[7] Lu, D., & Barber, A. H. (2012). Optimized nanoscale composite behaviour in limpet teeth. *Journal of the Royal Society, Interface*, *9*(71), 1318–1324.

[8] Wolff, M., Salikov, V., Antonyuk, S., Heinrich, S., & Schneider, G. A. (2014). Novel, highly-filled ceramic–polymer composites synthesized by a spouted bed spray granulation process. *Composites Science and Technology*, *90*, 154–159.

[9] Brandt, K., Wolff, M., Salikov, V., Heinrich, S., & Schneider, G. A. (2013). A novel method for a multi-level hierarchical composite with brick-and-mortar structure. *Scientific reports*, *3*, 2322.

[10] Georgopanos P., Eichner E., Filiz V., Handge U.A., Schneider G.A., Heinrich S., Abetz V., Improvement of mechanical properties by a polydopamine interface in highly filled hierarchical composites of titanium dioxide particles and poly(vinyl butyral), *Composites Science and Technology* 146 (2017) 73-82.

[11] Epstein, N., & Grace, J. R. (2011). Spouted and Spout-Fluid bed: Fundamentals and Application. Cambridge: Cambridge University Press.

[12] Eichner, E., Salikov, V., Bassen, P.-K., Heinrich, S., & Schneider, G. A. (2017). Using dilute spouting for fabrication of highly filled metalpolymer composite materials. *Powder Technology*. 316, 426–433.

[13] Rösler, J., Harders, H., & Bäker, M. (2012). *Mechanisches Verhalten der Werkstoffe.* (4th ed.). Wiesbaden: Springer Vieweg.

509541 Sedimentation and Magnetophoretic Velocity of Plain and Functionalized Magnetic Nanoparticles By in Situ Visualization of Separation Behavior in Superposed Gravity and Magnetic Fields (Invited)

Abstract:

Magnetic micro- and nanoparticles (MNP) have increasingly wide applications in information technology, magnetic fluids, selective separation, and nanomedicine. Their magnetic susceptibility (responsiveness) and magnetization has to be analysed during the particle design, optimization of application and quality control of production process. To this end, comprehensive characterization is demanded for both of the MP themselves and of the state of intermediate as well as final application ready dispersion.

The contribution introduces a recently developed measuring method characterizing magnetic MNP of particles and magnetic nanoparticles laden objects. Basically, gravity fields are superposed by parallel or orthogonal magnetic fields. Strength and direction of fields determine the acceleration of MNP and their direction of movement. Terminal velocity depends on field strength/gradient, buoyancy and hydrodynamic frictional coefficient of particles. Velocity can be determined with high precision and no assumption/material parameters by STEP-Technology® (1), recording instantaneously space and time resolved extinction profiles over the entire sample. Based on these profiles migration velocity is determined by software package SEPView®. A LUMiReader® adapted with permanent magnets of different field directions regarding gravity field was used. If cell holders with no magnets are used, particles can be analyzed regarding sedimentation velocity and particle size distribution as well as stability of the particle dispersion (1).

In a first study, magnets were positioned sidewise (orthogonal field) and applicability of the new approach investigated for a broad field of magnetic objects (MNP, assemblies of MNPs, siRNA-MNP, pDNA-MNP, virus-MNP, micro bubbles, labelled cells). Mean magnetophoretic velocity of analyzed particles amount to some tens μm/s up to over 100 μm/s.

On the other hand, we report in more detail on magnetophoretic mobility for monodisperse particles and magnetic viral complexes that were prepared via self-assembling of PEI-Mag2 or SO-Mag6-12 nanoparticles with adenoviral (Ad) or VSV-particles (VSV). We employed holders equipped with two disk neodymium-iron-boron magnets positioned underneath optical cell (magnetic flux density and gradient averaged over vertical sample height: 0.16 T and 33.5 T/m, respectively). Magnetophoretic velocity *Î½* was calculated based on integral normalized extinction, E/E_0, for different wavelengths, versus time *t* (2). No sedimentation of MNP was detected within time frame with no magnetic field application. A significant decline *of* E/E_0 (clarification) was observed with the magnetic field indicating a decrease of MNP concentration. From E/E_0 magnetophoretic velocity distribution was calculated for each wavelength. In case of monodisperse MNP magnetophoretic velocity distribution is independent on wavelength. In contrast, for polydisperse magnetic particles differences can be measured indicating that size classes have a different magnetic responsiveness (2). The newly developed STEPâ€MAG technology for in situ visualization and analysis of magnetophoretic velocity of MNP and micro-/ nanomagnetic assemblies has potential applications in design of advanced materials, molecule and cell separation, tissue engineering and drug/nucleic acid targeting.

[1] Lerche D., Comprehensive characterization of nano- and microparticles by in-situ visualization of particle movement using advanced sedimentation techniques. KONA Powder Particle J. (34) 2017 (open access)

[2] Mykhaylyk, O. et. al., Magnetophoretic mobility: Determination by space and time resolved extinction profiles (in-situ visualization). IEEE MAGNETICS LETTERS (6) 2015 (open access)

508496 Synthesis of Cerium Oxide Nanoparticles Under Reservoir-like Conditions (Invited)

Abstract:

Synthesis of cerium oxide nanoparticles under reservoir-like conditions

Pervaiz Shahid, Raza Ghulam, Amjad Muhammad, Lai Xiaojun and Wen Dongsheng

School of Chemical and Process Engineering, University of Leeds, Leeds, LS29JT

The energy demand of the world is rising and oil is presumed as a principal energy source for the next few decades. The era of finding â€œeasy oilâ€ is coming to an end and future supply will depend on fossil fuels supplied by non-conventional reservoirs and from enhanced oil recovery (EOR) methods. Around 50~70% of the original oil is still trapped in existing reservoirs after the primary and secondary recovery. Many EOR techniques have been developed based on thermal, chemical, mechanical and biological approaches. Recently, use of nanoparticles to improve oil production has received convincing interest in both academia and industry which greatly relies on the fabrication, stabilization and delivery of nanoparticles to the reservoir.

This project proposes a novel concept to synthesize nanoparticles *in-situ*, i.e., under reservoir-like conditions and to understand their formation kinetics. Different to conventional hydrothermal synthesis of nanoparticles, the uniqueness of harsh conditions (i.e. temperature up to 150°C, high pressure of a few hundred bars and high salinities, with the co-existence of oil/ gas phase) inside reservoirs require to understand the influence of temperature, pressure and salinity on nanoparticle production and their formation kinetics. This work focuses on the influence of these variables based on hydrothermal/ solvothermal reactions on size, morphology and stability of cerium oxide nanoparticles. The results revealed the influence of reaction temperature, pressure and salinity on the nanoparticle production and showed that stable nanoparticle dispersions, i.e. nanofluids, can be produced at a low temperature, high salinity and high pressure.

AUTHOR INDEX

AUTHOR INDEX

AUTHOR INDEX

AUTHOR INDEX